ENERGY POLICIES, POLITICS AND PRICES

BUSINESS CASE SCENARIOS FOR ENERGY EFFICIENCY

SELECT EQUIPMENT AND APPLIANCE SECTORS IN THE U.S., CHINA, AND INDIA

ENERGY POLICIES, POLITICS AND PRICES

Additional books in this series can be found on Nova's website under the Series tab.

Additional e-books in this series can be found on Nova's website under the e-book tab.

ENERGY POLICIES, POLITICS AND PRICES

BUSINESS CASE SCENARIOS FOR ENERGY EFFICIENCY

SELECT EQUIPMENT AND APPLIANCE SECTORS IN THE U.S., CHINA, AND INDIA

SEAN C. BYRNE
EDITOR

nova publishers
New York

For permission to use material from this book please contact us:
Telephone 631-231-7269; Fax 631-231-8175
Web Site: http://www.novapublishers.com

Library of Congress Cataloging-in-Publication Data

ISBN 978-1-62948-076-3

Published by Nova Science Publishers, Inc. † New York

CONTENTS

Preface		**vii**
Chapter 1	Business Case for Energy Efficiency in Support of Climate Change Mitigation, Economic and Societal Benefits in the United States *Michael A. McNeil, Nicholas Bojda, Jing Ke, Stephane de la Rue du Can, Virginie E. Letschert and James E. McMahon*	**1**
Chapter 2	Business Case for Energy Efficiency in Support of Climate Change Mitigation, Economic and Societal Benefits in China *Michael A. McNeil, Nicholas Bojda, Jing Ke, Yining Qin, Stephane de la Rue du Can, David Fridley, Virginie E. Letschert and James E. McMahon*	**77**
Chapter 3	Business Case for Energy Efficiency in Support of Climate Change Mitigation, Economic and Societal Benefits in India *Michael A. McNeil, Jing Ke, Stephane de la Rue du Can, Virginie E. Letschert and James E. McMahon*	**125**
Index		**165**

PREFACE

This book seeks to provide policymakers and other stakeholders with actionable information towards a road map for reducing energy consumption cost-effectively. A focus on individual end use equipment types (hereafter referred to as appliance groups) that might be the subject of policies - such as labels, energy performance standards, and incentives - to affect market transformation in the short term, and on high-efficiency technology options that are available today is examined. There is a strong, profit-based business case for investing in more energy-efficient products and designs. Energy efficiency, however, often is viewed as something that businesses and individuals "should" do as good citizens. The reality is that using energy inefficiently is like walking past money on the ground - money that could be put to far better use than paying electricity bills. Investing in energy efficiency therefore creates economic value. This book provides an overview of business case scenarios for energy efficiency in the United States, China and India.

Chapter 1 - This study seeks to provide policymakers and other stakeholders with actionable information towards a road map for reducing energy consumption cost-effectively. The authors focus on individual end use equipment types (hereafter referred to as *appliance groups*) that might be the subject of policies - such as labels, energy performance standards, and incentives - to affect market transformation in the short term, and on high-efficiency technology options that are available today.

There is a strong, profit-based business case for investing in more energy-efficient products and designs. Energy efficiency, however, often is viewed as something that businesses and individuals "should" do as good citizens. The reality is that using energy inefficiently is like walking past money on the ground - money that could be put to far better use than paying electricity bills. Investing in energy efficiency therefore creates economic value.

As the study title suggests, the high efficiency or *Business Case* scenario is constructed around a model of cost-effective efficiency improvement. The author's analysis demonstrates that a significant reduction in energy consumption and emissions is achievable at *net negative cost*, that is, as a profitable investment for consumers. Net savings are calculated assuming no additional costs to energy consumption such as carbon taxes. Savings calculated in this way relative to the base case are often referred to as "economic savings potential".

The United States energy demand picture is characterized by relatively high usage coupled with highly developed technology. Furthermore, the United States has a long history of appliance efficiency labels and standards. Therefore, efficiency improvement from U.S. appliances is characterized as successive stepwise improvement from an already efficient baseline. In spite of this already high baseline, the study finds significant opportunities for energy efficiency benefits at net negative cost.

Chapter 2 - This study seeks to provide policymakers and other stakeholders with actionable information towards a road map for reducing energy consumption cost-effectively. The authors focus on individual end use equipment types (hereafter referred to as *appliance groups*) that might be the subject of policies - such as labels, energy performance standards, and incentives - to affect market transformation in the short term, and on high-efficiency technology options that are available today.

As the study title suggests, the high efficiency or *Business Case* scenario is constructed around a model of cost-effective efficiency improvement. The authors' analysis demonstrates that a significant reduction in energy consumption and emissions is achievable at *net negative cost*, that is, as a profitable investment for consumers. Net savings are calculated assuming no additional costs to energy consumption such as carbon taxes. Savings relative to the base case as calculated in this way is often referred to as "economic savings potential".

Chinese energy demand has grown dramatically over the last few decades. While heavy industry still plays a dominant role in greenhouse gas emissions, demand from residential and commercial buildings has also seen rapid growth in percentage terms.

In the residential sector this growth is driven by internal migration from the countryside to cities. Meanwhile, income in both urban and rural subsectors allows ownership of major appliances. While residences are still relatively small by U.S. or European standards, nearly all households own a refrigerator, a television and an air conditioner. In the future, ownership rates are not expected to grow as much as in other developing countries, because they are already close to saturation. However, the gradual turnover of equipment in the world's largest consumer market provides a huge opportunity for greenhouse gas mitigation. In addition to residences, commercial floor space has expanded rapidly in recent years, and construction continues at a rapid pace. Growth in this sector means that commercial lighting and HVAC will play an increasingly important role in energy demand in China.

The outlook for efficiency improvement in China is encouraging, since the Chinese national and local governments have implemented significant policies to contain energy intensity and announced their intention to continue and accelerate these.

In particular, the Chinese appliance standards program, first established in 1989, was significantly strengthened and modernized after the passage of the Energy Conservation Law of 1997. Since then, the program has expanded to encompass over 30 equipment types (including motor vehicles). The current study suggests that, in spite of these efforts, there is significant savings to be captured through wide adoption of technologies already available on the Chinese market.

Chapter 3 - This study seeks to provide policymakers and other stakeholders with actionable information towards a road map for reducing energy consumption cost-effectively. The authors focus on individual end use equipment types (hereafter referred to as *appliance groups*) that might be the subject of policies - such as labels, energy performance standards, and incentives - to affect market transformation in the short term, and on high-efficiency technology options that are available today.

As the study title suggests, the high efficiency or *Business Case* scenario is constructed around a model of cost-effective efficiency improvement. The authors' analysis demonstrates that a significant reduction in energy consumption and emissions is achievable at *net negative cost*, that is, as a profitable investment for consumers savings are calculated assuming no additional costs to energy consumption such as carbon taxes. Savings relative to the base case as calculated in this way is often referred to as "economic savings potential".

The Indian energy demand picture is defined by very low average consumption that is growing very rapidly. For most end uses, this means that consumption in 2030 will be several times higher than it is today. Nearly half of Indian households are currently without electricity, and most do not own major appliances such as refrigerators and air conditioners. This situation is changing rapidly, however, with increased access to commercial energy sources and a swelling middle class with the purchase power to own and use major appliances and electronics. The low ownership/high growth situation means that the vast majority of equipment operating in 2030 will be installed after 2015. This means that efficiency measures implemented now will have maximal impact by the end of the forecast.

The organized commercial sector in India occupying large office buildings, retail outlets and hotels is still relatively small and consumes only a fraction of the energy as the household sector. Commercial building electricity consumption is growing, however, and savings there can be significant. Consumption is concentrated in lighting and air conditioning.

The industrial sector in India is a major energy consumer, with a large fraction of industrial electricity consumption passing through electric motors. Finally, losses in electricity distribution in India are high, partially due to a relatively inefficient stock of distribution transformers, which offer a significant opportunity for improvement.

Recognizing the need for enhanced energy efficiency in the face of exploding demand, the Government of India has already launched important programs to promote the adoption of high-efficiency equipment, including many actions already taken by the Bureau of Energy Efficiency (BEE) established as a direct consequence of the Energy Conservation Act of 2001. In a further important step, the Indian government announced its Action Plan on Climate Change (NAPCC) in June 2008, including creation of a National Mission on Enhanced Energy Efficiency. Finally, adoption of a proposed amendment to the Energy Conservation Act of 2001 would deepen the scope of government actions on energy efficiency.

So far, the Indian market has responded favorably to government efficiency initiatives, with Indian manufacturers producing a higher fraction of high-efficiency equipment than before program implementation. This study highlights both the financial benefit and the scope of potential impact for adopting this equipment, all of which is already readily available on the market.

In: Business Case Scenarios for Energy Efficiency ISBN: 978-1-62948-076-3
Editor: Sean C. Byrne

Chapter 1

BUSINESS CASE FOR ENERGY EFFICIENCY IN SUPPORT OF CLIMATE CHANGE MITIGATION, ECONOMIC AND SOCIETAL BENEFITS IN THE UNITED STATES*

Michael A. McNeil, Nicholas Bojda, Jing Ke, Stephane de la Rue du Can, Virginie E. Letschert and James E. McMahon

EXECUTIVE SUMMARY

This study seeks to provide policymakers and other stakeholders with actionable information towards a road map for reducing energy consumption cost-effectively. We focus on individual end use equipment types (hereafter referred to as *appliance groups*) that might be the subject of policies - such as labels, energy performance standards, and incentives - to affect market transformation in the short term, and on high-efficiency technology options that are available today.

There is a strong, profit-based business case for investing in more energy-efficient products and designs. Energy efficiency, however, often is viewed as something that businesses and individuals "should" do as good citizens. The

* This is an edited, reformatted and augmented version of the Berkeley National Laboratory, dated June 1, 2011.

reality is that using energy inefficiently is like walking past money on the ground - money that could be put to far better use than paying electricity bills. Investing in energy efficiency therefore creates economic value.

As the study title suggests, the high efficiency or *Business Case* scenario is constructed around a model of cost-effective efficiency improvement. Our analysis demonstrates that a significant reduction in energy consumption and emissions is achievable at *net negative cost*, that is, as a profitable investment for consumers. Net savings are calculated assuming no additional costs to energy consumption such as carbon taxes. Savings calculated in this way relative to the base case are often referred to as "economic savings potential".

The United States energy demand picture is characterized by relatively high usage coupled with highly developed technology. Furthermore, the United States has a long history of appliance efficiency labels and standards. Therefore, efficiency improvement from U.S. appliances is characterized as successive stepwise improvement from an already efficient baseline. In spite of this already high baseline, the study finds significant opportunities for energy efficiency benefits at net negative cost. These include:

Final Energy savings:

- 230 billion kWh of electricity and 0.14 EJ of natural gas per year in 2020
- 430 billion kWh of electricity and 0.35 EJ of natural gas per year in 2030
- A total of 4900 billion kWh of electricity and 3.1 EJ natural gas cumulatively through 2030

The end uses studied in this report are concentrated in the buildings sector, and the bulk of the savings is to be found in electricity savings. The study finds that by 2030, cost-effective efficiency improvements could save 430 billion TWh, which represents about 12% of total buildings electricity consumption in that year, according to the International Energy Agency's World Energy Outlook[1]. This corresponds to 9.3% of electricity consumption in the U.S.

Primary Energy Savings:

- 1.3 EJ of coal per year in 2020 and 0.58 EJ of natural gas per year in 2020

- 2.4 EJ of coal per year in 2030 and 1.2 EJ of natural gas per year in 2030
- A total of 27 EJ of coal and 13 EJ of natural gas cumulatively through 2030

This corresponds to 9.3% of electricity consumption in the U.S. or 3.7% of total energy consumption, in primary energy terms.

Cumulative emissions mitigation:

- 3000 million metric tons of CO_2 through 2030
- 12 million metric tons of SO_2 through 2030
- 3.5 million metric tons of NO through 2030
- 62 metric tons of mercury through 2030

The *Business Case* saves 265 mt CO_2 in 2030. According to WEO, with current policies in place, emissions in the United States will total 5310 mt CO_2 or 712 mt more than the Annex I target of 4598 mt (5.2% below the 1990 level of 4850 mt.) Implementation of the *Business Case* would close over a third of this gap.

Financial impacts to consumers through 2030:

- Equipment investment of 260 billion dollars (USD)
- Energy bill savings of 560 billion dollars (USD)
- Net savings of 300 billion dollars (USD)

In addition to the 300 billion dollars net savings provided to consumers through 2030, annual savings of 430 billion kWh would avoid capital investments of 32 billion dollars[2].

Job Creation:

- Net creation of 85,000 to 200,000 jobs

The approach of the study is to assess the impact of short-term actions on long-term impacts. "Short-term" market transformation is assumed to occur by 2015, while "long-term" energy demand reduction impacts are assessed in 2030. In the intervening years, most but not all of the equipment studied will turn over completely. The 15-year time frame is significant for many products, in the sense that delay of implementation postpones economic benefits and mitigation of emissions of carbon dioxide. Such delays would result in putting in place energy-wasting technologies, postponing improvement until the end of their service life, or potentially resulting in expensive investment either in additional energy supplies or in early replacement to achieve future energy or emissions reduction targets.

The *Business Case* concentrates on technologies for which cost-effectiveness can be clearly demonstrated. The appliance groups studied are:

Residential Equipment:

Incandescent Lamps Refrigerators
Room Air Conditioners Water Heaters Televisions
Standby Power
Electric Cooking Equipment
Central Air Conditioners and Heat Pumps
Gas Furnaces

Commercial Equipment:

Air Conditioners and Heat Pumps
Commercial Linear Fluorescent Ballasts
High-Intensity Discharge Lamps
Distribution Transformers
Boilers

Energy savings and greenhouse gas emissions mitigation for these appliance groups are summarized in Table ES-1

Since the study includes only appliance groups for which cost-effectiveness can be clearly demonstrated, the benefits determined represent only a subset of the economy-wide potential. Specifically, transportation end uses and industrial processes technologies are not covered, because data sufficient to include them were not possible to collect within the scope of the research. Likewise, the study does not include system approaches such as smart grids. These approaches to efficiency may have important impacts but the calculation of costs and benefits is not as straightforward as for individual pieces of equipment. In addition, the technologies analyzed represent a snapshot of what is currently on the market. Technological innovations are certain to occur over the coming decades, and these will likely present new opportunities for efficiency improvement, and exert downward pressure on costs.

Efficiency measures are determined to be cost-effective if the cost of conserved energy associated with them is less than the consumer's energy price, that is, the amount saved in cumulative energy bills is greater than the initial investment. The *Business Case* scenario is generated by identifying the *maximum efficiency improvement for which cost of conserved energy is lower than utility energy prices* (projected to 2015). The relative contribution to cumulative emissions for each appliance group is shown in Figure ES-1.

Table ES-1. Energy Savings and Pollutant Mitigation by Appliance Group

Appliance Group	Final Energy Savings			Emissions Mitigation					
	In 2020	In 2030	Through 2030	In 2020	In 2030	Through 2030			
	TWh			mt CO_2		mt CO_2	mt SO_2	mt NO	t Hg
Electric Water Heaters	41.8	92.4	887	24.3	53.1	512	2.0	0.7	10.4
Incandescent Lamps	31.9	0.0	861	18.6	0.0	513	1.9	0.7	10.1
Distribution Transformers	33.3	89.8	764	19.4	51.6	441	1.7	0.6	9.0
Commercial Lighting	31.0	74.3	679	18.0	42.7	392	1.5	0.6	8.0
Furnaces	21.4	62.0	512	4.3	12.5	103	0.0	0.0	0.0
Standby Power	24.5	43.9	463	14.3	25.2	268	1.0	0.4	5.4
Central AC & HP	21.6	49.8	457	12.6	28.6	264	1.0	0.0	5.4
Refrigerators	15.7	39.3	348	9.1	22.6	201	0.8	0.1	4.1
Televisions	20.2	22.6	300	11.8	13.0	174	0.7	0.2	3.5
Gas Water Heaters	13.2	27.2	269	2.7	5.5	54.3	0.0	0.0	0.0
Room AC	5.5	9.2	99.2	3.2	5.3	57.4	0.2	0.1	1.2
Commercial Boilers	3.2	7.5	68.9	0.6	1.5	13.9	0.0	0.0	0.0
Commercial AC & HP	2.0	5.0	45.2	1.2	2.9	26.1	1.0	0.0	5.4
Electric Cooking Equipment	0.1	0.2	1.3	0.1	0.1	0.8	0.003	0.001	0.0
Total	**265**	**523**	**5754**	**140**	**265**	**3021**	**11.9**	**3.5**	**62**

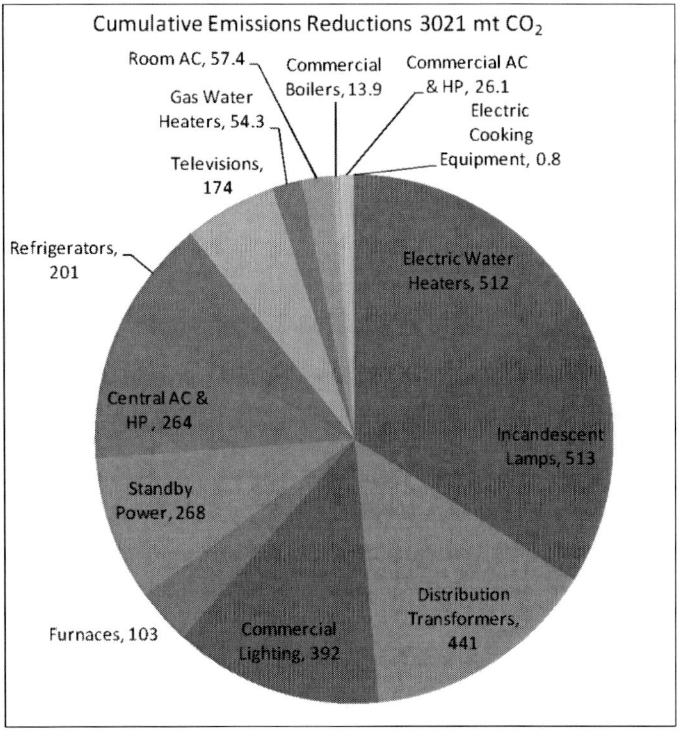

Figure ES-1. Contribution to Cumulative CO_2 Emissions Reductions 2010-2030.

Several conclusions can be drawn from Table ES-1 and Figure ES-1. First, the largest accessible savings appears to be available in the residential sector. Notable among these are electric water heaters, for which heat pump technology was found to be cost-effective[3]. Incandescent lamps also show a large savings through a phase out and replacement with CFLs[4]. The Energy Independence and Security Act passed by Congress in 2007 includes a phase out of incandescent lamps starting in 2012 which will capture this savings when implemented. Large potential savings are also forecast for televisions, for which a minimum efficiency standard has never been set, but is under consideration. Similar potential is available for refrigerators, which have been regulated with several updates, but for which high-technology options such as vacuum insulation panels are just now becoming cost-effective. Finally, standby power reduction shows high potential savings. It is the policy of DOE to address standby power as part of future standards for electronics and other products with standby modes.

Savings are relatively low for commercial buildings' end uses, with the exception of lighting. While the dominant technology (fluorescent tube lamp ballasts) are already relatively efficient, much of the savings in this category comes from high intensity discharge lamps which, while constituting a much smaller fraction of the end use, permit large percentage savings, possibly as a result of not having been regulated in the past. Some of the limitation on savings is due to recent standards for HVAC equipment that restrict further cost-effective improvement given the available data[5]. Furthermore, an ever-growing fraction of commercial building energy is used in miscellaneous products for which efficiency improvement may be possible, but for which data are not yet available. Lastly, distribution transformers do show significant savings because of the scale of distribution losses at the national level, even though total efficiency improvement is small on a percentage basis, due in part to the recent standards rulemaking.

Table ES-2. Cumulative Financial Impacts of Efficiency Improvement through 2030

Appliance Group	Cumulative Financial Impacts				
	Cost	Savings	Net	NPV @ 3% DR	NPV @ 7% DR
	$ Billions				
Incandescent Lamps	15.9	95.1	**79.2**	66.4	53.2
Commercial Lighting	9.4	65.3	**55.9**	36.2	21.2
Standby Power	9.3	51.2	**41.9**	27.5	16.3
Distribution Transformers	41.4	73.5	**32.0**	20.7	12.0
Televisions	6.0	33.1	**27.1**	18.4	11.3
Electric Water Heater	71.4	97.9	**26.5**	17.2	10.1
Central AC & HP	35.0	50.5	**15.4**	10.0	5.9
Refrigerator	31.3	38.4	**7.1**	4.6	2.7
Room AC	6.1	11.0	**4.9**	3.2	1.9
Gas Water Heaters	8.9	12.1	**3.2**	2.1	1.2
Furnaces	20.0	23.0	**3.0**	2.0	1.1
Commercial Boilers	1.0	2.7	**1.7**	1.1	0.6
Commercial AC & HP	3.5	4.3	**0.8**	0.5	0.3
Electric Cooking Equipment	0.1	0.1	**0.1**	0.04	0.02
Total	**259**	**558**	**299**	**210**	**138**

The analysis shows that cost-effective efficiency improvement could yield very significant financial benefits to U.S. consumers. Table ES-2 shows positive net savings for all appliance groups, which is not surprising, since the target efficiency levels were constructed to be cost- effective. The table shows that cost-effective efficiency improvements require an investment of 259 billion dollars over the next 20 years, but these investments will return over twice as much over the same period, for a net savings of 299 billion dollars, or of order of one thousand dollars per capita. The present value of net savings is 210 billion dollars assuming a discount rate of 3%, and 138 billion dollars with a 7% discount rate. Of the appliance groups studied, electric water heaters require the largest investment at $71.4 billion, but also have the highest payoff at $97.9 billion. Phasing out incandescent lamps has a similar payoff of $95.1 billion, but requires an investment of only $15.9 billion. This technology has the highest net benefit, of $79.2 billion, with commercial lighting providing the second biggest, with $55.9 billion.

1. INTRODUCTION

This study seeks to provide policymakers and other stakeholders with actionable information towards a road map for reducing energy consumption in the most cost-effective way. A major difference between the current study and some others is that we focus on individual equipment types that might be the subject of policies - such as labels, energy performance standards, and incentives - to affect market transformation in the short term, and on high-efficiency technology options that are available today.

The approach of the study is to assess the impact of short-term actions on long-term impacts. “Short term” market transformation is assumed to occur by 2015, while “long-term” energy demand reduction impacts are assessed in 2030. In the intervening years, most but not all of the equipment studied will turn over completely. The 15-year time frame is significant for many products however, indicating that delay of implementation postpones impacts such as net economic savings and mitigation of emissions of carbon dioxide. Such delays would result in putting in place energy-wasting technologies, postponing improvement until the end of their service life, or potentially resulting in expensive investment either in additional energy supplies or in early replacement to achieve future energy or emissions reduction targets.

1.1. Description of Efficiency Measures to Date

The United States energy demand picture is characterized by relatively high usage coupled with highly developed technology. The commercial /service sector is very well-developed, and energy use in commercial buildings is correspondingly high, with total greenhouse gas emissions from this sector expected to overtake those of the residential sector within the next 20 years. Because of cold winters and warm summers in large regions, both space heating and space cooling are major end uses, and natural gas used in space and water heating constitutes a significant source of carbon dioxide emissions.

Ownership of major appliances (refrigerators, water heaters, televisions, washing machines and air conditioners) is nearly universal in U.S. households and while the capacity of some appliances continues to grow, energy consumption for the traditional major end uses is growing slowly. Most future growth in residential energy consumption is therefore expected to be due to consumer electronics, small appliances, standby power and other "plug loads".

Furthermore, the United States has a long history of appliance efficiency standards. Therefore, efficiency improvement from U.S. appliances is characterized as successive stepwise improvement from an already efficient baseline. While there remain opportunities for significant savings, some further major improvements may depend on innovative technologies which have not been commercialized and for which market prices are not yet known.

1.2. Policies and Programs to Encourage Efficiency

Regulatory measures such as the implementation of a minimum energy performance standard (MEPS) for new equipment coming on the market were first introduced in the U.S. after the first oil crisis of 1974. In addition, instruments such as labels, awards, and public information campaigns are used to increase consumers' knowledge and awareness of energy efficiency. Finally, financial incentives address the market barrier of high upfront costs of energy-efficient products.

In 1975 the Energy Policy and Conservation Act (EPCA) established the Energy Conservation Program for Consumer Products Other than Automobiles. This program started the first generation of appliance standards, focusing on residential end-uses. In 1978 the National Energy Conservation Policy Act amended the EPCA, changing the standards from voluntary to mandatory and requiring updates of the pre-existing standards. As DOE's

Energy Efficiency and Renewable Energy Office reports, standards were not well developed until the 1980s[6]. The National Appliance Energy Conservation Policy Act (NAECA) of 1987 amended EPCA and established MEPS for household appliances, as well as revision schedules for DOE. The residential end-uses analyzed in this report had their first mandatory legislation start with NAECA.

Standards for commercial products were introduced with the Energy Policy Act of 1992 (EPAct 1992). A second Energy Policy Act was issued in 2005 (EPAct 2005), which introduced or renewed standards for over a dozen end-uses, covering commercial industrial and residential sectors. The EPAct of 1992 and 2005 together address most of the commercial end-uses covered in this study. The most recent major legislative action is the Energy Independence and Security Act of 2007 (EISA).

From 1987 to date, U.S. appliance standards have been implemented on 47 product types (ASAP, 2011). Today DOE is conducting more rulemakings than at any time in its history. Appendix 1 shows a list of equipment for which standards have been implemented at the federal level or state level. The list shows the date by which new standards were first implemented, the date they were issued by DOE or Congress, the date they became or will become effective and the date for scheduled updates. This information is collected by Appliance Standards Awareness Project (ASAP, 2011).

In parallel to MEPS, the U.S. Environmental Protection Agency (EPA) established an energy consumption labeling scheme starting in 1991 to promote energy-efficient product purchases and reduce greenhouse gas emissions. Computers and monitors were the first labeled products. Through 1995, EPA expanded the label to additional office and residential equipment. In 1996, the EPA partnered with the U.S. Department of Energy to expand significantly the range of equipment covered with the Energy Star label. Today, more than 40,000 individual product models, produced by nearly 3,000 manufacturers across more than 60 product categories.have earned the ENERGY STAR label. The ENERGY STAR program is estimated to have saved 204 billion kWh in 2009, about 5 percent of U.S. electricity demand (EPA, 2010). This amounts to nearly $17 billion from customer utility bills.

Other forms of energy efficiency programs include financial incentive measures that are implemented by utilities or independent state agencies. In the U.S., the power sector is decentralized and regulation is done at the state level. About half of the states have passed legislation to apply a small levy or public benefit charge as a fraction of a cent per kWh on electricity sales to finance a common public fund that is then used to support energy efficiency programs. The revenue from this charge is redistributed to consumers in the form of financial incentive programs.

Public benefits charges range from $0.00003 to $0.003 per kWh with a median value of about $0.0011 per kWh. Utility spending on energy efficiency represents between 0.7% and 3% of total utility retail revenue (Kushler, 2004). According to the last CEE report on the "State of the Efficiency Program Industry" (Caracino, 2010), this generates a total annual budget of $5.4 billion in 2010 for rate-funded electric efficiency programs across all States, an increase of 44% compared to 2009. Electric efficiency programs in the residential sector 31% of the budget in 2010. California, New York, Florida, and Massachusetts account for 50% (or $2.7 billion) of the total amount budgeted for electric energy efficiency for 2010. California alone accounts for 1.5 billion (or 28%) of which $436 million is directed to residential sector energy efficiency improvements. Electricity program budgets per capita vary widely across US states, ranging from 54.8 dollars in Vermont to 3 cents in Virginia (Caracino, 2010). In terms of energy savings, the CEE reports that combined U.S. electricity rate funded projects saved about 92,578 GWh in 2009, of which 26,876 GWh were in the residential sector (29%).

California has a long history in implementing utility demand-side management (DSM) programs and established itself as a leader in promoting energy efficiency in the U.S. Utilities started to implement utility DSM programs in the early 70's in California. A recent study conducted by the California Energy Commission (CEC, 2005) shows annual savings of 40,000 GWh and 12,000 megawatts of peak electricity, equivalent to 24 500-megawatt power plants (CEC, 2005).

2. Energy Demand Scenarios

As the study title suggests, the high efficiency or *Business Case* scenario is constructed around a model of cost-effective efficiency improvement. The point of the study is to demonstrate that a significant reduction in energy consumption and emissions is achievable at a net negative cost, that is, as a profitable investment for society. There are a variety of ways of assessing costs and benefits to society. We chose to focus on the end user's perspective: costs in terms of additional retail equipment prices (capital investments); savings from reduced energy bills (operating costs). Only direct energy savings are included, without valuing non-energy benefits that may also accrue (comfort, productivity, health). Finally, the cost-benefit analysis is made without the elevated effective energy prices that could be implied by carbon taxes, carbon trading schemes or other policies. Savings relative to the base case as calculated in this way is often referred to as "economic savings potential".

A national-level high-efficiency scenario is constructed by assuming that market transformation to high-efficiency technologies will occur by 2015, which is judged to constitute the "short term" by the study, because it considers that five years is sufficient time to achieve market transformation through aggressive policies and stakeholder actions. The study does not model specific actions, which could include mandatory standards, voluntary labeling programs, voluntary agreements by manufacturers, utility demand-side-management programs and others[7].

The target efficiency level chosen is that which *maximizes efficiency while providing a net benefit to consumers*. This is to be contrasted with scenarios which maximize consumer payoff but not necessarily efficiency improvement, or those that include the best available technology ("max tech") without consideration of cost-effectiveness. Consumer cost-benefit analysis is evaluated in terms of cost of conserved energy. Cost of conserved energy (CCE) is the amortized incremental cost of equipment divided by annual energy savings. In other words, it's the additional annual capital investment needed to purchase high-efficiency equipment instead of baseline equipment, divided by the energy savings provided by the investment. This quantity, which has units of dollar per unit energy, can be compared to prevailing energy prices to assess consumer cost-effectiveness. Technologies with a CCE less than forecast energy prices in 2015 are deemed cost-effective.

A few comments about whether this definition is optimistic or pessimistic are warranted. On one hand, high efficiency technologies are compared to the current baseline technology, even though there may already be a market for higher efficiency equipment, and the average efficiency of the market is constantly improving. This tends to underestimate the baseline forecast and overestimate savings. On the other hand, it likely underestimates the efficiency that will be achievable in a cost-effective way, first of all because technology costs are generally decreasing (according to technological learning rates) and the emergence of new technologies that may not be available for analysis. Therefore, there are two compensating effects not taken into account in the analysis. The results should therefore be taken as representative of the scale of potential improvement, not as a reliable prediction. The methodology is chosen to maximize concreteness and defensibility by relying on technologies that can be justified by actual cost data.

2.1. Literature Review

Some recent examples of studies that have identified potential energy savings from energy efficiency improvements include:

China

- China's appliance standards are estimated to have saved 1.08 EJ during 2006-2008, with refrigerators, air conditioners and televisions contributing the bulk of the savings. (Price, L, et al, Energy Policy (in press), 2011. http://china.lbl.gov/sites/china.lbl.gov/files/ACEStudy.2011.pdf
- (Fridley 2008) estimates potential savings of 1.2 TWh in 2012 and 16 TWh by 2020 for energy labels on refrigerators in China. Fridley, D., Zheng, N., Zhou, N., Aden, N., Lin, J., Chen, J., and Sakamoto, T. 2008. *China Refrigerator Information Label: Specification Development and Potential Impact.* LBNL-246E, LBNL, Berkeley, CA. http://china.lbl.gov/sites/china.lbl.gov/files/LBNL_246E._China_Refrigerator_Information_ Label._Feb2008.pdf
- (Cheung 2008) describe the growth of China's energy efficiency industry, projecting spending of USD 300 billion over five years. Cheung, R., and Kang, A. 2008. *China's Booming Energy Efficiency Industry.* World Resources Institute (WRI). http://pdf.wri.org/chinas_booming_energy_efficiency_industry.pdf
- (Aden 2010) uses lifecycle assessment to show that for buildings in the Beijing area, 80% of energy use and related emissions is due to operations, and about 20% due to materials. Aden, N.,Qin, Y., and Fridley, D., 2010. *Lifecycle Assessment of Beijing-Area Building Energy Use and Emissions: Summary Findings and Policy Applications.* LBNL, Berkeley, CA. http://china.lbl.gov/sites/china.lbl.gov/files/LBNL-3939E.pdf
- (Zhou 2010) provides an overview of China's policies on energy efficiency. Zhou, N., Levine, M.D., and Price, L., 2010. "Overview of Current Energy Efficiency Policies in China," *Energy Policy.* http://china.lbl.gov/sites/china.lbl.gov/files/Overview.Energy_Policy_November2010.pdf

India

- (Delio 2009) estimates potential savings from energy efficiency across all sectors in India to be 183 TWh in five years. Delio, E.A., Lall, S., and Singh, C., 2009. *Powering Up: The Investment Potential of Energy Service Companies in India.* World Resources Institute. http://pdf.wri.org/powering_up_full_report.pdf
- (De la Rue du Can 2009) provides both retrospective and prospective views of energy use in the residential and transport sectors of India. de la Rue du Can, S., Letschert, V., McNeil, M., Zhou, N., and Sathaye, J., 2009. *Residential and Transport Energy Use in India: Past Trend and Future Outlook.* LBNL, Berkeley, CA. http://ies.lbl.gov/drupal.files/ies.lbl.gov.sandbox/LBNL-1753E.pdf

United States

- The National Research Council report, *America's Energy Future,* in 2009 estimated potential cost-effective energy savings in the U.S. of about 20% in 2020 and about 30% in 2030, with the greatest potential in the buildings sector (National Research Council, *Limiting the Magnitude of Future Climate Change*, 2010).
- The American Physical Society report, *Energy Future: Think Efficiency* (2008) estimated 572 TWh of electricity savings in the residential sector in 2030, and about 30% savings for the building sector as a whole, all below the retail price of electricity.
- The U.S. Department of Energy's Appliance Standards Programs has conducted extensive studies for regulated product types (http://www1.eere.energy.gov/buildings/appliance_standards/), identifying economically justified and technologically feasible energy efficiency improvements.
- The Energy Information Administration annually publishes additional efficiency scenarios, e.g., high technology cases, in conjunction with the *Annual Energy Outlook* (http://www.eia.doe.gov/oiaf/aeo/).

2.2. Construction of the Energy Demand Scenarios

Any study that aims to project energy efficiency improvements from specific technologies must make the link between unit-level improvements and national impacts. The current study achieves this using LBNL's Bottom-Up Energy Analysis System (BUENAS). As the name suggests, BUENAS is a bottom-up technology-oriented model, rather than a top-down macroeconomic model[8]. BUENAS combines unit-level efficiency scenarios with a forecast of stock size and turnover to calculate national energy savings impacts through 2030. Unit level energy demand by baseline and "target" technologies are collected in a database that the model takes as inputs, and which define the base case and high efficiency scenarios. Growth of the stock (number of units operating) by 2030 is a function of economic and population growth.

BUENAS uses minimum efficiency performance standards (MEPS) as a default policy, that is, it models a discrete change in the efficiency of equipment after a specific year. For the current study, we chose an implementation year of 2015, assuming that several years lead time are necessary between identification of efficiency targets, and making them mandatory.

Originally constructed as a global model, BUENAS covers a wide range of energy-consuming products, including most appliance groups generally covered by Energy Efficiency Standards and Labeling (EES&L) programs around the world. The global model covered the following appliance groups:

- *Residential Sector:* Lighting, Refrigerators, Air Conditioners, Fans, Washing Machines, Standby Power, Televisions, Electric Ovens, Space Heating and Water Heating.
- *Commercial Building Sector:* Lighting, Air Conditioning, Refrigeration, Ventilation, Office Products, Space Heating and Water Heating.
- *Industrial Sector:* Electric Motors.

For the purposes of the *U.S. Business Case for Energy Efficiency*, many of the end uses needed for the analysis were present in BUENAS. However, many modifications were made. First, the *Business Case* model is dependent on an evaluation of cost-effectiveness. Therefore, appliance groups for which data were insufficient to permit this calculation were not included. On the other hand, some equipment types for which data were available were not included in the original model. In that case, these end uses were added. Finally, the efficiency scenario originally constructed for the U.S. in BUENAS was updated in light of recent DOE rulemakings. In some cases, an updated standard meant that further improvement was not cost-effective (or data do not exist), so the end use was dropped. In other cases, the rulemaking resulted in data that were not available at the time the original model was built. In those cases, both the baseline and target level were updated. Several omissions are notable. First, commercial furnaces provide about half of the space heating in the commercial building sector, no data were available for this product. Furnaces were covered in (USDOE 2010c), but not analyzed for cost-effectiveness. Likewise, no data for commercial water heaters were provided, so this product was omitted.

Finally, industrial motors are an important end use that are generally found to provide cost- effective improvement opportunities. The Energy Policy Act of 1992 (EPACT 1992) established nominal efficiency levels for electric motors. Most recently, this standard was updated in the Energy Independence and Security Act of 2007. This new standard, which makes mandatory a voluntary standard ("NEMA Premium"), covers electric motors from 1 to 500 hp, and came into effect in 2010. DOE is currently considering an update to the rulemaking set forth in the Energy Independence and Security Act of 2007. Technical studies supporting this rulemaking were not available at the time of the current study; therefore, motors were not included.

The BUENAS model uses the Long Range Energy Alternatives Planning (LEAP) platform[9] to forecast energy consumption by end use from 2005 (base year) to 2030. The strategy of the model is to first forecast end use *activity*, which is driven by increased ownership of household appliances and growth in the industrial sector. The total stock of appliances can be modeled either according to an econometric diffusion model or according to unit sales forecasts, if available. Electricity consumption or *intensity* of the appliance stock is then calculated according to estimates of the baseline intensity of the prevailing technology in the local market. Finally, the total final energy consumption of the stock is calculated by modeling the flow of products into the stock and the marginal intensity of purchased units, either as additions or as replacements of old units. The high efficiency or "policy" scenario is created by the assumption of increased unit efficiency relative to the baseline starting in a certain year. For example, if the average baseline unit energy consumption (UEC) of new refrigerators is 450 kWh/year, but a MEPS taking effect in 2012 requires a maximum UEC of 350 kWh/year, the stock energy in the policy scenario will gradually become lower than that of the base case scenario due to increasing penetration of high-efficiency units under the standard. By 2030, the entire stock will generally be impacted by the standard[10]. Figure 1 shows the analytical structure of BUENAS.

The main outputs of BUENAS are base case energy consumption forecasts to 2030 by end use and energy, energy saving impacts of the modeled policy, and carbon dioxide emissions mitigation impacts. For this study, financial impacts were added to the model in a spreadsheet calculation.

For the residential sector, *activity* as modeled in Module 1 of the model is given by the stock of equipment, that is, the number of appliances installed and operating in U.S. households in a given year. Three different methods are used to estimate the total stock of a particular residential end use. For each region and end use, the method is chosen which is deemed to produce the most accurate result, and for which sufficient data are available. In order of priority, the methods are:

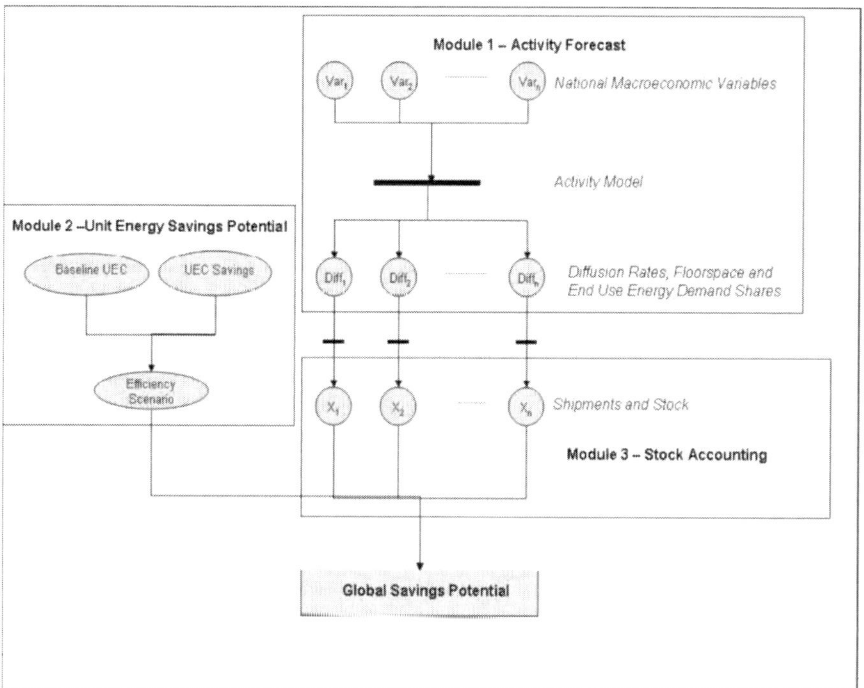

Figure 1. Structure of BUENAS.

Method 1 - Stock based on historical and projected flows of products (unit sales).

Method 2 - Stock from historical and projected ownership rates – sales derived from stock increases and replacement rates.

Method 3 - Stock from econometric modeling driven by macroeconomic trends – sales derived from stock increases and replacement rates.

Sales forecasts are available as part of the U.S. rulemaking analysis for residential products. Therefore, the model was updated in most cases to model stock according to the first method listed above, which represents an improvement over the other methods in most cases.

Once the number of residential products in each appliance group in each year is established, this number is multiplied by the *annual unit energy consumption* (UEC) to yield energy demand for the appliance group. *UEC* is the subject of Module 2 of BUENAS, and determines the efficiency scenario modeled. Determination of the baseline and efficiency scenario UEC is discussed in Section 4 below.

Finally, Module 3 tracks the introduction of each year's cohort of appliances into the stock, taking account of growth in the market, equipment retirements, and replacements. Retirement and survival functions are derived from average lifetimes and assumed to have a distribution around the mean value. This shape of the retirement function is assumed to be that of a normal distribution centered around the mean lifetime by default, but takes the form of a more complicated function (Weibull distribution) if such a distribution is available. The survival function is given by:

$$Survival\,(age) = 1 - \int Retirement(age)$$

Using the retirement distribution, the model calculates the weighted average efficiency of the stock in each year. In the case of the high efficiency scenario, only a small fraction of the stock operates at high efficiency in the years immediately following the policy start date, but this fraction grows over time. The percentage of stock operating in 2030 that was installed after the policy start date is dependent on the assumed average lifetime of the product class.

Compared to the residential sector, energy demand in the commercial building sector is driven by a much wider variety of equipment types and follows distinct usage patterns depending on the type of building. For this reason, BUENAS models commercial buildings in an aggregate fashion, rather than at the level of individual appliances. The activity variable in this case is commercial building floor space. Commercial floor space projections for the United States are taken from DOE's Annual Energy Outlook 2010 (EIA 2010), which takes the place of Module 1 for the commercial sector. In Module 2, the commercial sector model uses aggregate energy intensity numbers for major appliance categories, such as lighting, space heating and air conditioning and refrigeration. Commercial clothes washers and distribution transformers were added to the model as individual appliances according to shipments forecasts. In order to model energy demand and savings from efficiency improvement, we estimate the fraction of energy covered by individual technologies for which data are available. Energy and demand are thereby calculated from base year values of energy intensity according to a scaling factor.

3. Efficiency Improvement Potential – Cost-Benefit Analysis

The primary sources for appliance group details in the United States are publicly available technical studies performed by the U.S. Department of Energy supporting its residential and commercial appliance standards program. These studies generally include forecasts of equipment stock and sales, baseline efficiency consumption, and use patterns. In addition, they provide an assessment of the price to consumers of equipment incorporating energy efficiency improvement measures and the resulting performance. These "cost efficiency curves" are the basis for determination of cost-effective efficiency targets incorporated in this study. Secondary sources are used for products not covered by DOE standards.

Because cost-efficiency data for the United States are generally taken from published results of appliance standards rulemakings either completed or in process, some interpretation of market baselines is needed. In the case where data are available but rulemakings are not finalized (preliminary analysis) we assume the same baseline as the DOE analysis (usually the previous standard). In cases where the available data corresponds to an already implemented standard, we adjust the baseline to the new standard.

Cost effectiveness is defined in terms of cost of conserved energy, that is, how much the end user must pay in terms of annualized incremental equipment investment for each unit of energy saved by higher efficiency equipment. The formula for cost of conserved energy is

$$CCE = \frac{I \times q}{S} \qquad \text{Eq. 1}$$

In this equation, I is the total additional investment needed to purchase high efficiency equipment rather than the baseline technology, and S is the resulting annual energy savings. The capital recovery factor q is given in turn by:

$$q = \frac{d}{(1-(1+d)^{-L})} \qquad \text{Eq. 2}$$

In this equation, *d* is the end user discount rate and *L* is the average lifetime of the equipment, in years. Defined in this way, *I* times *q* is an annual payment for an amortized capital investment. Cost of conserved energy is a convenient metric for comparison of cost-effectiveness of measures[11].

3.1. Construction of Cost vs. Efficiency Curves

Over the past decade, technical parameters permitting efficiency cost-benefit analysis of a significant number of appliances[12] have become available as a result of analyses performed in support of the U.S. Department of Energy's Appliance Standards Program[13]. In fact, the appliances for which data are available now account for the majority of energy consumed in the U.S. residential sector. The data provided to support DOE rulemakings forms the basis of our cost-benefit calculation for many of the products analyzed. Obviously, using data supporting regulations in process brings up the question of additionality of savings capture, since much of the savings we will forecast is already the subject of DOEs program. Indeed, it is not our intention to suggest that the savings we present is being "left on the table", but only to say that the opportunities for cost-effective improvement are significant, and some of these opportunities will certainly be captured through aggressive policies like the DOE program. To be as precise as possible about what is considered still "on the table" and what is already part of the baseline, we followed the following guidelines:

- Rulemakings for which a final rule has not been issued were not considered part of the *Base Case*. The baseline efficiency level used is the current baseline, as estimated by DOE.
- Rulemakings for which a final rule has been issued and for which standards come into effect before 2015 *are* considered as part of the *Base Case*. In this case, we adjusted the baseline to correspond to the new standard. Efficiency improvements in the *Business Case* are possible, if found to be cost-effective relative to the adjusted baseline.
- Rulemakings for which a final rule has been issued but comes into effect in 2015 or beyond *are* included in the *Business Case*. Because the *Business Case* considers energy efficiency policy taking effect in

2015, we include DOE standards set for that year or beyond as part of the *Business Case*, but adjust the target level according to the cost-effectiveness criterion. There is one appliance group that falls into this category (residential water heaters).

As a requirement of the DOE rulemaking process (NECPA, Pub.L. 95-619, 92 Stat. 3206, 42 U.S.C.), DOE must consider impacts on U.S. consumers and manufacturers. DOE typically analyzes costs and benefits likely to be result from any proposed efficiency rulemaking for appliances. The root of this assessment is an engineering analysis of possible efficiency improvements and their costs, and a national impact analysis which calculates the impacts of replacing commonly used equipment components with more efficient ones. The output of this analysis is a set of efficiency levels (called "design options" in the TSD), each of which is associated with reduction of energy consumption and increase in manufacturing cost. The relationship between design option costs and resulting energy consumption (or efficiency) is often referred to as a *cost-efficiency curve*. The cost-efficiency curve forms the basis of consumer cost-benefit analysis of efficiency improvement regulations. The parameters sourced from DOE Technical Support Documents (*TSDs*) needed to establish cost-efficiency curves are:

- *Baseline Unit Energy Consumption* – The annual electricity or gas energy consumption used by "standard efficiency" products. Energy consumption is generally estimated assuming typical use patterns. The standard efficiency product may represent the minimum efficiency required by the standard in force, or may be somewhat higher according to the market at the time of the analysis.
- *Design Option Unit Energy Consumption* – Annual electricity or gas consumption for each efficiency improvement technology, assuming no change in use patterns.
- *Baseline Equipment Price* – Estimated retail price of "standard efficiency" products. Baseline Equipment Prices are generally calibrated to actual prevailing retail prices. Equipment Prices may include the cost of installation, but do not include maintenance and repair costs[14]. Equipment prices given in DOE reports were inflation-adjusted using GDP deflators provided by (BEA, 2010).

- *Design Option Equipment Price* – Projected price of equipment for each efficiency improvement technology. Design option equipment prices are generally calculated by adding estimated production costs, and then applying estimates of manufacturer and retail markups.

DOE categorizes covered products into separate product classes and formulates a separate energy conservation standard for each product class. The criteria for separation into different classes are type of energy used, capacity, and other performance-related features. Each product class has features that determine the above parameters in a distinct way. Therefore, cost-efficiency curves are determined at the product class level, and a cost-benefit calculation is performed on each individually. Efficiency and cost parameters are then calculated at the appliance class level by weighting according to market shares. For simplicity, we consider only those product classes that account for two percent of the market or more as estimated in the TSD. This selection generally covers the large majority of the market and we extend the results of these to the entire appliance group.

In order to demonstrate the construction of the cost-efficiency curve, we present the example of one product class. We chose refrigerator freezers with top mounted freezer compartments and auto-defrost features as a demonstrative example. This product class is the most common refrigerator on the U.S. market, representing 54.3% of refrigerator sales. Price and energy consumption are based on 18 ft^3 total volume. Baseline technology parameters, efficiency design options and resulting efficiency improvement are given in Table 1.

Starting from the baseline configuration, alternate designs are constructed by replacing or adding components in turn, in order of cost-effectiveness. Each of these options is represented by a point in Figure 2. The resulting *cost-efficiency curve* has a typical shape with increasing costs per unit efficiency improvement.

The cost-benefit evaluation is a calculation that determines whether increases in first cost are justified by a return on investment in the form of reduced utility bills and accounting for discounting of delayed costs or savings. Using the cost of conserved energy metric, a cost- effective efficiency improvement costs less per unit energy saved than the energy tariff, with the initial investment amortized over the life of the appliance. Appliance groups and product classes are described in Section 3.2. The data used in this analysis are detailed in the tables in Appendix 2. Sources for these tables are summarized in the tables in Appendix 3. The following parameters are collected or calculated for each product class:

Table 1. Efficiency Design Options for Top-Mount Refrigerator-Freezers

Efficiency Level	Design Option
Baseline	See Table 5-A.2.1 of 1
10%	Increase Condenser Size by 100% & Increase Compressor EER from 5.55 to 6.1
15%	Increase Compressor EER from 6.1 to 6.26 & Use Brushless DC Condenser Fan Motor
20%	Increase Evaporator Size by 14% & Use Adaptive Defrost & Use Variable Speed Compressor
25%	12.2 sqft VIP in FZR Cabinet
30%	2.9 sqft VIP in FZR Door & 7.1 sqft VIP in FF Door & 6.7 sqft VIP in FF Cabinet
30.60%	1.9 sqft more VIP in FF Cabinet

Source: Table 5-A.3.1 of USDOE (2010a).

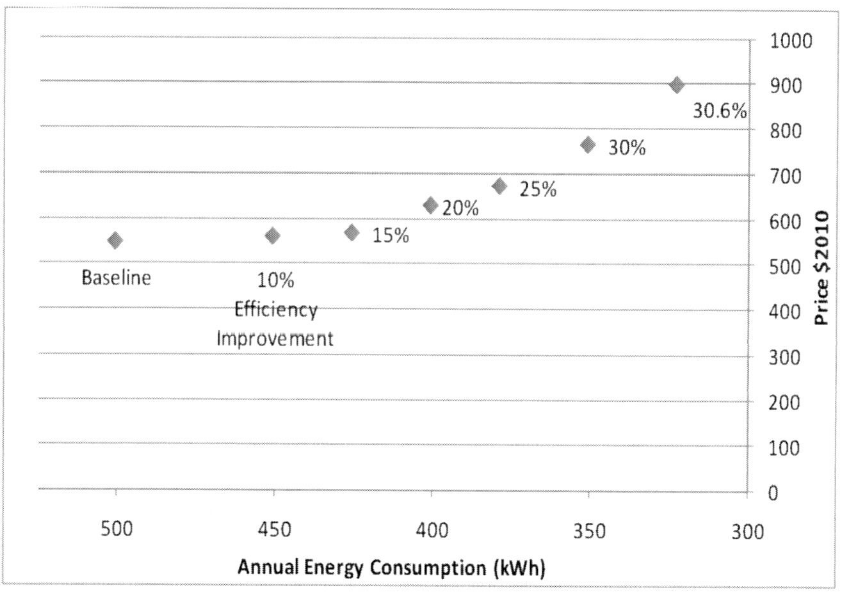

Source: Table A.2.1.

Figure 2. Cost vs. Efficiency for Top-Mount Refrigerator-Freezers.

3.2. Equipment Data

Incandescent Lamps

Incandescent lamps constitute the bulk of residential lighting energy. According to the Building Energy Data Book (BEDB 2010)[15], lighting accounts for 14.2% of residential electricity or 10% of residential building energy, in primary energy terms. Replacement of incandescent lamps with compact fluorescent lamps (CFLs) or other technology such as LEDs is generally at the top of the list of attractive efficiency measures because of the large fractional savings (up to 60%) and the high degree of cost-effectiveness. Bans on incandescent lamps are also among the most popular efficiency policies globally. The United States has a planned phase-out of incandescent bulbs in 2012 as specified in the Energy Independence and Security Act of 2007. Although this regulation is currently "on the books", we chose to include its effects as part of the *Business Case* scenario, rather than in the Base Case. In this case, the high-efficiency scenario is characterized by successful implementation and enforcement of the policy, rather than the establishment of it. Cost-effectiveness of incandescent lamp replacement is not highly controversial. For simplicity, we assume a retail cost of $5.00 for a 15 W CFL that lasts for 5 years versus $2.50 for five incandescent lamps, which we assume to last only a year each. Even with these conservative assumptions, incandescent lamp replacement is among the most cost- effective measures available.

Refrigerators

Taken together, refrigerators and freezers account for 8.9% of residential electricity or 6.3% of primary residential energy consumption in the United States (BEDB 2010), making them the highest consumption end use aside from lighting, water heating and HVAC (EIA 2010 Table 4). Efficiency improvement design options for refrigerators include added and/or improved insulation, increased compressor efficiency and increased evaporator and condenser area (see Table 1). As part of its upcoming final rule for updated refrigerator standard (USDOE 2010a) to take effect in 2014, DOE has released data showing equipment costs and energy consumption for the 7 product classes covered by the rulemaking. We analyzed the three product classes with more than 2% of the market. These three refrigerator–freezer classes; top mounted freezer, bottom-mounted freezer and side-by side, represent about 97% of the current market. Cost and energy consumption parameters for refrigerator products are given in Table A.2.1.

Room Air Conditioners

Space cooling accounts for 22.4% of residential electricity or 15.8% of U.S. residential energy use in primary terms (BEDB 2010). While the dominant technology for residential space cooling in the United States is central air conditioning, room air conditioners still play an important role, and as they have recently experienced a dramatic drop in price, the market for them has grown in residences that previously had no air conditioning, and in some small commercial buildings. Window units are still dominant in the United States, unlike much of the world, where the trend is toward split system RACs. Room air conditioner efficiency can be improved through a variety of design improvements (see USDOE 2010b). The analysis takes advantage of preliminary analysis supporting a rulemaking to be published in early 2011[16] with an implementation date of 2014. We consider four separate product classes, defined by cooling capacity and the presence of louvers. Cost and energy consumption parameters for refrigerator products are given in Table A.2.2.

Water Heaters

Electric and fuel-burning water heaters account for 8.9% of residential electricity and 27.4% of residential natural gas, or 13.2% of residential primary energy consumption (BEDB 2010). DOE passed new standards for residential water heaters in 2010. These standards will come into effect in 2015. Therefore, we consider the current baseline in construction of the *Base Case*, and construct the *Business Case* according to the cost-benefit criterion. The resulting target levels may or may not correspond to the standard set by DOE. Electric storage tank water heaters enjoy a sizeable market share in the United States. They have been subject to several standards updates by DOE and their efficiency is close to 100%. Since there is no virtually no waste heat from the electric heating element, remaining heating efficiency of conventional units is limited to reducing standby losses through improved insulation around the tank. However, over the past decade water heaters using heat pump technology have been introduced into the U.S. market[17]. These units can produce hot water with less than half of the energy required by traditional water heaters, and represent the next major leap in efficiency. Like electric water heaters, gas storage tank water heater efficiency can be improved by reducing standby losses through added tank insulation. In addition, efficiency can be improved by increasing heat transfer efficiency, usually by increasing the size of heat exchangers. Very high efficiency gas water heaters use *condensing* technology in which the flue gases are cooled to a degree to where they form a condensate. Cost and energy consumption parameters for water heaters are given in Table A.2.3.

Televisions

Televisions are estimated by the BUENAS model to consume 67 TWh of electricity which corresponds to 4.6% of residential electricity or 3.2% of total residential primary energy as estimated by BEDB 2010[18]. Televisions are a dynamic and rapidly evolving technology. Recent market trends include a massive shift to flat panel technology, with dramatic increases in screen size, along with market-driven efficiency improvements. Because of the dynamism in television technology, efficiency baselines and technology trends are difficult to forecast and historical cost data are not relevant. Recent history has shown, however, that efficiency targets such as Energy Star have been easily met, often capturing most of the market by the date they are implemented with virtually no increase in price. For this reason, we judged it a reasonable assumption that Energy Star 4.0 announced in 2010 will be the baseline by 2015, and that Energy Star 5.0 will be attainable through cost-effective technology[19]. We assume a cost of conserved energy of two cents per kWh for improvement from Energy Star 4.0 to 5.0 for televisions.

Standby Power

Standby power is estimated by the BUENAS model to consume 34 TWh of electricity which corresponds to 2.3% of residential electricity or 1.6% of total residential primary energy as estimated by BEDB 2010[20]. Standby power consumption is a feature of a wide range of products, including major appliances, consumer electronics and home entertainment equipment. This mode of power consumption is increasingly shown to be a major source of energy demand, and has become a prominent candidate for efficiency improvement (IEA, 2001). Reduction of standby power is typically very inexpensive to achieve through redesign of electronic components. DOE has decided not to consider standby power in a single rulemaking for all products, but will include the reduction of standby power as part of individual rulemakings governing active mode use for each product type. For simplicity, we include standby power as an aggregate category, and assume an 80% reduction in standby power per product, from 5W to 1 W, at a cost of two cents per kWh.

Electric cooking appliances (Electric)

Electric cooking appliances account for 2.2% of residential electricity or 1.5% of residential primary energy consumption (BEDB 2010). Five electric cooking appliances were considered in DOE's recent rulemaking announced in 2009 (USDOE 2009). These are: coil cooktops, smooth cooktops, standard ovens, self-cleaning ovens and microwave ovens. Cost and energy consumption parameters for cooking appliances are given in Table A.2.4.

Central Air Conditioners and Heat Pumps

Space cooling accounts for 22.4% of residential electricity or 15.8% of U.S. residential energy use, in primary terms (BEDB 2010).Mechanical cooling is also the major contributor to peak electricity demand in the U.S.. Ducted central air conditioners, which are by far the most common means of cooling, were first regulated by NAECA in 1987. A major update of standards came into effect in 2006, with the next expected to come into effect in 2015. Space heating through electric heat pumps is also common in some regions of the United States. Due to the similarity of CAC and HP technologies, they are regulated together as product classes of the same appliance group. Cost and energy consumption parameters for central air conditioner and heat pump product classes are given in Table A.2.5

Gas Furnaces

Natural gas space heating contributes 67.1% of residential natural gas or 21.1% to residential primary energy demand (BEDB 2010). Natural gas warm-air furnaces are the most common form of space heating in U.S. homes. The main efficiency improvement options for furnaces are related to improvement of heat transfer, to minimize the amount of waste heat escaping through the flue. Like water heaters, high-efficiency gas furnaces use condensing technology. Residential furnace and boiler standards were updated in 2007 and are set to come into effect in 2015. We do not consider further improvement for boilers because furnaces are much more common. For furnaces, however, recent standards were set at the current market baseline. We therefore do consider cost-effective improvement options from this baseline. Cost and energy consumption parameters for gas furnace product classes are given in Table A.2.6.

Commercial Air Conditioners and Heat Pumps

Together with space heating and lighting, air conditioning is one of the most important commercial building end uses. Space cooling accounts for 12.6% of commercial sector electricity or 10.1% of total commercial building primary energy (BEDB 2010). New standards for commercial central air conditioning and heat pumps came into effect in 2010. We analyzed the potential for further cost-effective improvement from this new baseline. Cost and energy consumption parameters for commercial air conditioners and heat pumps are given in Table A.2.7.

Commercial Linear Fluorescent Ballasts

Lighting accounts for 22.2% of commercial sector electricity or 17.4% of total commercial building primary energy (BEDB 2010). According to DOE's *Commercial Building Energy Consumption Survey* (EIA, 2003), linear fluorescent lamps account for roughly 75% of lighting in commercial environments. While this type of lighting is already highly efficient, improvements are still possible, and even small percentage improvements can have a large effect on overall sector energy consumption. A standards rulemaking issued in 2000 taking effect in 2005 required electronic fluorescent lamp ballasts. The next rulemaking is expected in 2011 and will come into effect in 2014. Our analysis is based on data made available in the preliminary stage of the upcoming rulemaking. Cost and energy consumption parameters for commercial linear fluorescent lamp ballasts are given in Table A.2.8.

Commercial High Intensity Discharge Lamps

Lighting accounts for 22.2% of commercial sector electricity or 17.4% of total commercial building primary energy (BEDB 2010). HID lamps produce light by creating an electrical arc between electrodes in a tube that is filled with gas and metal salts. HID lamps are used in applications which include street lighting, security lighting, and lower wattage applications. To date, HID have never been regulated at the federal level. Recently, however, a DOE determination found significant saving potential for this class of lighting. Therefore, even though this lighting type accounts for only about 10% of commercial sector lighting, the potential for savings, in percentage terms, could be significant. Cost and energy consumption parameters for high intensity discharge lamps are given in Table A.2.9.

Distribution Transformers

Distribution Transformers are designed to reduce the voltage of the electricity coming from the electric grid, to a lower voltage which is applicable for appliances and other electricity- consuming systems. Losses in transformers account for an inefficiency in national electricity usage on the order of a percent. Distribution transformers efficiency is currently regulated by standards that came into effect in 2007 and 2010. New standards are due in 2012. We analyzed distribution transformers for cost-effective improvement using the data from the previous rulemaking and adjusted the baseline to account for the recent standards. Cost and energy consumption parameters for distribution transformers are given in Table A.2.10.

Commercial Boilers

Space heating accounts for 3.8% of commercial sector electricity and 51.9% of commercial sector natural gas, or 13.4% of total commercial building primary energy (BEDB 2010). Boilers provide roughly half of space heating in commercial buildings. Commercial boilers are regulated by a standard announced in 2009 that comes into effect in 2012. This standard covered efficiency requirements for several products included in voluntary standards issued by the American Society of Heating, Refrigerating and Air-Conditioning Engineers (ASHRAE). DOE made the decision to make mandatory requirements of ASHRAE's ASHRAE 90.1 2007 standard. The analysis used data generated for that rulemaking. Data were not available for commercial furnaces. Cost and energy consumption parameters for high intensity discharge lamps are given in Table A.2.11.

3.3. Cost of Conserved Energy Calculation

As mentioned in the previous section, annual unit energy consumption (*UEC*) and equipment price (*Price*) are shown for all product classes considered in the analysis in Appendix 2, unless otherwise specified[21]. These parameters are used in the calculation of cost of conserved energy according to Equation 1 by comparing each design option to the baseline, according to:

$$I = Price_{DesignOption} - Price_{Baseline}$$

and

$$S = UEC_{Baseline} - UEC_{DesignOption}$$

The parameters used in calculation of q in Equation 2 are as follows:

Product Lifetime (L) – Average number of years that a product is used before failure and retirement. Lifetimes vary by product class and are estimated from manufacturer reports, or from survey data.

Table 2. Cost of Conserved Energy Calculation for Refrigerators

Class	Top Mount Auto Defrost Refrigerator -Freezer			Bottom-Mount Refrigerator-Freezer			Side-by-Side Refrigerator-Freezer		
Market share	54.3%			13.5%			28.9%		
Lifetime	17.1			17.1			17.1		
q	0.088			0.088			0.088		
Efficiency Level	UEC	Price	CCE	UEC	Price	CCE	UEC	Price	CCE
	kWh/yr	$2010	$2010	kWh/yr	$2010	$2010	kWh/yr	$2010	$2010
Baseline	**501**	**547**		**613**	**953**		**768**	**1161**	
1	451	559	0.021	551	955	0.003	691	1164	0.003
2	426	567	0.024	521	957	0.004	653	1169	0.006
3	401	629	0.072	490	963	0.007	614	1188	0.015
4	379	672	0.091	459	1028	0.043	576	1254	0.042
5	351	765	0.128	429	1136	0.087	538	1396	0.090
6	323	899	0.175	391	1286	0.132	515	1508	0.120
Target Annual Incremental Cost (I×q)	$11			$16			$21		
Target UEC (kWh/yr)	379			429			538		

Weighted Average Baseline UEC	588 kWh/yr
Weighted Average Target UEC	428 kWh/yr
Weighted Average Energy Savings	161 kWh/yr
Weighted Average Incremental Cost	14.5 $2010
Product CCE	0.090 $2010

Discount Rate – Discount rates vary somewhat from product to product in DOE's analysis. For simplicity, we assume a consumer discount rate of 5% for residential products and 6% for commercial products, values that are similar to

DOE estimates, which range from 4.0% to 6.2% for residential products and from 4.2% to 7.0% for commercial products. The consumer discount rate we use is a "real" discount rate, and represents the actual rate of interest on the financing of the equipment, the mortgage rate if new appliances are included in the home purchase.

Using these parameters, we calculate cost of conserved energy for each design option for each product class. The results of this calculation, shown in the Appendix 2 tables, are the basis of construction of the efficiency scenario.

As stated above, the target efficiency level chosen is that which *maximizes efficiency while providing a net benefit to consumers*. Following this definition, we identify the target UEC for each product class as the lowest UEC for which cost of conserved energy is below the utility price. In order to model efficiency improvements beginning in the year 2015, we use average national residential and commercial prices[22] provided by DOE's Annual Energy Outlook 2010 (EIA 2010). The average residential electricity price is $0.11 per kilowatt-hour ($/kWh) and the average residential natural gas price is $13.17 per million British thermal units ($/MMBtu). The corresponding energy prices in the commercial sector are $0.10/kWh and $11.40/MMBtu, respectively.

To illustrate the construction of the efficiency scenario, we return to the refrigerator example. The top portion of Table 2 shows *UEC*, *Price* and *CCE* for each design option of the three major product classes. The lower portion of the table shows the target annual incremental cost and UEC for each class. For example, design options 1 through 4 can be implemented at a cost of $0.09/kWh or less, which is less than the electricity price of $0.10/kWh, while the cost of implementing design options 5 and 6 exceed the price of electricity. Therefore we determine design option 4 to be the maximum cost-effective technology. This design option costs $125 more to implement than the baseline, which gives an annual incremental cost of $11 when multiplied by q. The UEC for this design option is 379 kWh/yr, so the savings is 122 kWh with respect to the baseline.

Following this pattern with the other product classes, we find a weighted average energy savings of 161 kWh per year and weighted average incremental cost of $14 per year, for a weighted average CCE of 14/161 or $0.09/kWh which is less than the utility price by construction[23]. Appendix 2 shows the calculation of CCE for each product class, and calculates weighted average target UEC and CCE for each appliance group. The cost of conserved energy for all appliance groups is compared to utility prices in Figure 3.

The main inputs to the construction of the two scenarios, the *Base Case* and the *Business Case* scenario are the baseline *UEC* and the UEC established by CCE in Figure 3. We call this the *Business Case UEC*. Baseline UEC, *Business Case* UEC, Percent Decrease and Cost of Conserved Energy are presented in Table 3.

Table 3 shows that energy efficiency improvements can be made to a wide variety of equipment that will provide not only energy savings, but financial benefits to consumers. It also demonstrates the importance of performing this type of analysis at the appliance group level, since the cost-effective potential varies widely between appliance groups.

We emphasize that the results may not match exactly with findings of the recent DOE rulemakings. There are several possible reasons for that:

1. The analysis presented here is highly simplified compared to the DOE analysis, which takes into account distributions of household characteristics, regional differences, marginal energy prices, etc.
2. The DOE analysis bases its evaluation of cost-effectiveness not only at the individual level (*Life-Cycle Cost Analysis*) but also from national financial impacts (*Net Present Value*).
3. The DOE analysis takes into consideration other factors, such as feasibility in ramping-up production of high efficiency technology to a mass scale.

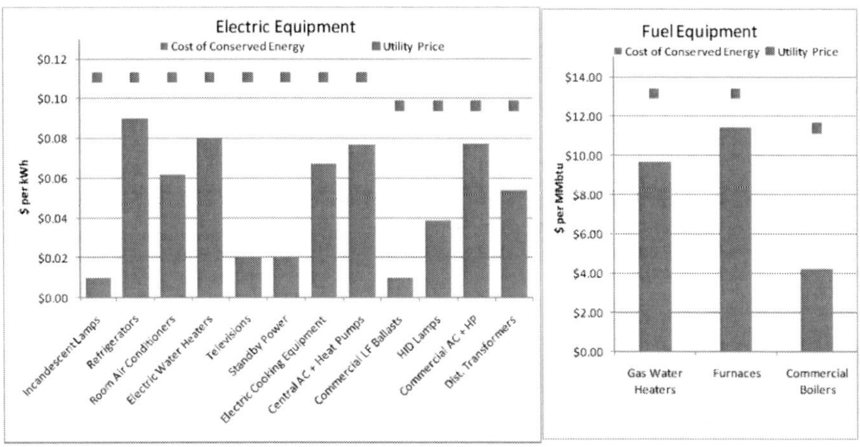

Figure 3. Cost of Conserved Energy and Energy Prices.

Efficiency improvements found to be cost-effective for each appliance group are as follows:

Incandescent Lamps – Not surprisingly, switching from incandescent lamps to CFLs is found to be highly cost-effective. Switching to LEDs was not considered due to lack of data demonstrating significant cost reductions for LED technology.

Table 3. Annual Energy Use Per Unit and Cost of Conserved Energy

Equipment	Baseline UEC	Business Case UEC	Percent Decrease	Cost of Conserved Energy
Electric Equipment	**kWh/a**	**kWh/a**	**%**	**$/kWh**
Commercial Linear Fluorescent Ballasts	306	285	7%	$0.010
Incandescent Lamps	46.5	15.1	67%	$0.018
Televisions*	200	141	30%	$0.020
Standby Power**	17.2	3.4	80%	$0.020
HID Lamps	729	335	54%	$0.038
Distribution Transformers (Losses Only)	5702	2720	52%	$0.054
Room Air Conditioners	700	581	17%	$0.062
Electric Cooking Equipment	153	152	1%	$0.067
Central AC + Heat Pumps	3285	2525	23%	$0.077
Commercial AC + HP	8901	8608	3%	$0.078
Electric Water Heaters	2604	1216	53%	$0.081
Refrigerator	588	428	27%	$0.090
Fuel Equipment	**MMBtu/a**	**MMBtu/a**	**%**	**$/MMBtu**
Commercial Boilers	2106	2042	3%	$4.241
Gas Water Heaters	16.6	14.7	11%	$9.68
Furnaces***	52.1	48.2	8%	$11.42

* Results shown for LCD display only.

** Per product. Standby power is calibrated to total estimated household standby power.

*** Residential boilers excluded due to recent rulemaking.

Refrigerators – Despite dramatic improvements to this appliance in the past, the analysis finds that there are still affordable improvements to be made, notably through the use of vacuum insulated panels.

Room Air Conditioners – Design improvements to this appliance are also still available, but savings is not dramatic in the U.S. context.

Water Heaters – The analysis finds heat pump technology to be cost-effective, producing a large percentage savings for this high-intensity end use[24]. The analysis finds condensing technology for water heaters to be cost-effective.

Televisions – A 30% improvement is assumed based on Energy Star requirements, and assumed to be cost-effective.

Standby Power – A reduction in standby power from 5W to 1W per product is assumed to be cost-effective.

Electric Cooking Equipment – Savings from design options in this appliance group were found to be small, and often not cost-effective, so the weighted average improvement is very small.

Central Air Conditioning and Heat Pumps – Like refrigerators, cooling is one of the most highly-regulated residential products. Additional cost-effective savings potential is likely the result of decreases in the cost of high-efficiency compressor technology in the last decade.

Residential Furnaces – A rulemaking in 2007 covered both residential furnaces and boilers, but an update is currently in progress for furnaces only. Our analysis finds that a shift to condensing gas furnaces would be cost-effective to consumers on an average basis.

Commercial Linear Fluorescent Ballasts – Due to the high intensity of this technology in commercial buildings, additional cost-effective savings potential are available. Improvement of this technology was found not to be cost-effective in the residential sector.

High Intensity Discharge Lamps – High intensity discharge lamps have not yet been regulated by the DOE standards program, but a recent DOE determination found significant energy savings potential for this product, which triggers a future rulemaking proceeding. Our analysis also finds significant cost-effective efficiency improvement potential across all product classes considered in this category.

Distribution Transformers – Revised standards for distribution transformers came into effect in 2010. Our readjustment of the baseline to the new standard results in the finding that additional cost-effective savings is possible, given our assumptions.

Commercial Boilers – DOE issued a rulemaking in 2009 covering commercial boilers along with other equipment covered by the ASHRAE 90.1 2007 voluntary standard. DOE subsequently decided to set standards for boilers at the ASHRAE level, to come into effect in 2012. Although these standard levels are included in our Base Case, we find significant (7%) additional cost- effective improvement.

Commercial Air Conditioners and Heat Pumps – The most recent standards for these product classes came into effect in 2010. Using the data supporting the previous rulemaking, we readjusted the baseline to correspond to the current standard. Using this method, we find that additional cost-effective potential is possible but, not surprisingly, the magnitude is small compared to savings from the pre-standard baseline.

4. NATIONAL LEVEL ENERGY SAVINGS OPPORTUNITIES

Because of the modular structure of the BUENAS model (see Figure 1), once the inputs are established it is a relatively straightforward process to construct the two energy demand scenarios and compare them to calculate savings potential. The full details of the calculation of energy demand are provided in (McNeil 2008) and are omitted here. Instead, we present a summary of inputs modified and added to the model for the present study.

Activity was largely modeled through Method 1 described in Section 2.2. Sales forecasts for most products were taken directly from the TSD documents or accompanying analysis tools downloaded from the DOE website without modification. These forecasts are generally highly detailed and complex, and are based on relevant drivers, including housing construction, normal replacements, early replacements and fuel switching. The exceptions to this method of activity forecasting are as follows:

Incandescent Lamps – Incandescent lamps are considered in the residential sector only. Total points of light in 2002 are based on a recent study (IEA 2006) and grown according to the relationship between number of light fixtures and household income as determined by (McNeil 2008). In the *Base Case*, incandescent are assumed to be completely phased out in favor of CFLs by 2020. The *Business Case* is based on the scenario set forth in EISA, which phases out incandescent in the two years following 2012.

Televisions – Television ownership is calibrated to data provided by DOE's *Residential Energy Consumption Survey 2005* (EIA 2005). Since U.S. television ownership rates are already the highest in the world, we assume no further increase in the number of televisions per household. The stock of TVs therefore grows only with the number of U.S. households. Market shares of CRT, LCD and Plasma display technologies and average screen size are taken from (Display Search 2010).

Standby Power – BUENAS models the total wattage of standby power per household according to a model determined in (McNeil 2008). This model was calibrated to a recent report (IEA 2001) which found an average of 67 W per U.S. household in 2005.

Commercial Lighting – Commercial lighting intensity in the United States in terms of kWh/m^2 is modeled in (McNeil 2008). In order to calculate energy demand and savings in the *Business Case*, we consider the percentage savings found to be cost-effective for linear fluorescent ballasts and high-intensity discharge lamps shown in Table 3 and consider their estimated contribution to commercial building lighting energy according to DOE's *Commercial Building Energy Consumption Survey* (EIA 2003) to derive a weighted average percentage improvement, which is applied to the *Business Case* lighting intensity.

Commercial Heating and Air Conditioning Equipment – Commercial air-source air conditioners and heat pump savings was applied to overall cooling intensity according to market shares in a similar way to commercial lighting. Commercial boiler savings was also applied to heating in this way. Since only about 40% of heating is accomplished with boilers, overall heating savings is relatively small.

4.1. Energy Savings and Emissions Reductions

Site energy savings is the basis for all national impacts calculations. Site energy demand refers to electricity and natural gas consumed in a home or business, and does not include fuel inputs in generation of electricity, or losses in transmission or distribution. Site energy is the energy affected most immediately by efficiency improvement. It is also the energy consumption that appears on consumer utility bills, and forms the basis for the cost-benefit analysis detailed above.

Site energy consumption is calculated by BUENAS for both the *Base Case* and *Business Case* scenarios. Energy *activity* is the same in both cases[25], so the difference between them is driven by the trend in *marginal intensity*, that is, the UEC of products sold in each year. The UEC for the two scenarios are identical until the policy implementation date of 2015[26]. After that date, the efficiency target in the *Business Case* is the high efficiency level determined by cost-benefit analysis, while it remains at the baseline efficiency level in the *Base Case*. The difference in UEC in the two scenarios applies only to new products – in this way, the policy modeled has the structure of a minimum efficiency performance standard, and does not imply retrofits of existing equipment. By 2016 overall energy demand of stock in the *Business Case* is only slightly lower than the *Base Case*, because only one year's sales are affected by the policy. Moving through the forecast, LEAP tracks the gradual flow of high efficiency products into the stock and the retirement of less efficient ones, so that the average stock UEC gets closer to the target level. Depending on the lifetime of the product, the entire stock may not be converted by 2030, since some low-efficiency products installed before 2015 will survive. Figure 4 shows the evolution of site energy savings by appliance group. From 2012-2015, all of the savings is due to the accelerated phase out of incandescent lamps in the *Business Case*. This savings goes to zero by 2020 however, as incandescent lamps are eliminated by this date in both scenarios. From 2015 onward, energy savings is growing for all other products as high efficiency products begin to penetrate the stock in the *Business Case*.

Site energy savings results are summarized in Table 4. Total savings for all appliance groups totals 523 TWh in the year 2030. Cumulative savings through 2030 total 5754 TWh.

Table 4. Energy Savings and Pollutant Mitigation by Appliance Group

Appliance Group	Final Energy Savings			Emissions Mitigation					
	In 2020	In 2030	Through 2030	In 2020	In 2030	Through 2030			
	TWh			mt CO_2		mt CO_2	mt SO_2	mt NO	t Hg
Electric Water Heaters	41.8	92.4	887	24.3	53.1	512	2.0	0.7	10.4
Incandescent Lamps	31.9	0.0	861	18.6	0.0	513	1.9	0.7	10.1
Distribution Transformers	33.3	89.8	764	19.4	51.6	441	1.7	0.6	9.0
Commercial Lighting	31.0	74.3	679	18.0	42.7	392	1.5	0.6	8.0
Furnaces	21.4	62.0	512	4.3	12.5	103	0.0	0.0	0.0
Standby Power	24.5	43.9	463	14.3	25.2	268	1.0	0.4	5.4
Central AC & HP	21.6	49.8	457	12.6	28.6	264	1.0	0.0	5.4
Refrigerators	15.7	39.3	348	9.1	22.6	201	0.8	0.1	4.1
Televisions	20.2	22.6	300	11.8	13.0	174	0.7	0.2	3.5
Gas Water Heaters	13.2	27.2	269	2.7	5.5	54.3	0.0	0.0	0.0
Room AC	5.5	9.2	99.2	3.2	5.3	57.4	0.2	0.1	1.2
Commercial Boilers	3.2	7.5	68.9	0.6	1.5	13.9	0.0	0.0	0.0
Commercial AC & HP	2.0	5.0	45.2	1.2	2.9	26.1	1.0	0.0	5.4
Electric Cooking Equipment	0.1	0.2	1.3	0.1	0.1	0.8	0.003	0.001	0.0
Total	**265**	**523**	**5754**	**140**	**265**	**3021**	**11.9**	**3.5**	**62**

Emissions reductions are calculated directly from energy savings according to a carbon factor. The carbon factor for electricity includes fuel inputs to generation, and accounts for transmission and distributions losses. The carbon factor is taken from (EIA 2010) and is 0.594 kg CO2/kWh in 2015 decreasing to 0.574 kg CO2/kWh in 2030. Carbon factors for natural gas and fuel oil are assumed to remain constant at 0.202 and 0.264 kg/CO2, respectively. Emissions reductions from energy savings determined by multiplying energy savings by carbon factors are shown in Table 4. Total mitigation in the *Business Case* is found to be 265 mt CO_2 in 2030 and 3021 mt CO_2 over the entire forecast. Figure 5 shows the contribution to cumulative CO_2 mitigation from all appliance groups. In addition, we also evaluate the amount of other pollutants avoided by energy savings, including SO2, NO and mercury (Hg). Avoided pollutants are estimated according to emissions factors available in (EIA 2010)

Several conclusions can be drawn from Table 4 and Figure 5. First, the largest accessible savings appears to be available in the residential sector. Notable among these are electric water heaters, for which heat pump technology was found to be cost-effective[27]. Incandescent lamps also show a large savings through a phase-out and replacement with CFLs[28]. The Energy Independence and Security Act passed by Congress in 2007 includes a phase out of incandescent lamps starting in 2012 which will capture this savings when implemented. Large potential savings is also forecast for televisions, for which a minimum efficiency standard has never been set, but is in process. Similar potential is available for refrigerators, which have been regulated with several updates, but for which high-technology options such as vacuum insulation panels are just now becoming cost-effective. Finally, standby power reduction shows high potential savings. It is the policy of DOE to address standby power as part of future standards for electronics and other products with standby modes.

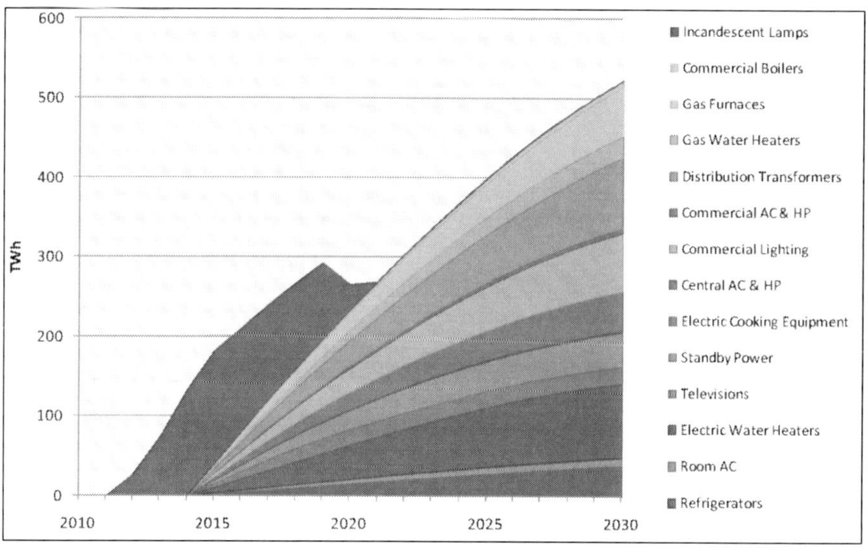

Figure 4. Site Energy Savings by Appliance Group– 2010-2020.

Savings are relatively low for commercial buildings end uses, with the exception of lighting. While the dominant technology (fluorescent tube lamp ballasts) are already relatively efficient, much of the savings in this category comes from high intensity discharge lamps which, while constituting a much smaller fraction of the end use, permit large percentage savings, possibly as a result of not having been regulated in the past. Some of this is due to recent standards for HVAC equipment that restrict further cost-effective improvement given the available data[29]. Furthermore, an ever-growing fraction of commercial building energy is used in miscellaneous products for which efficiency improvement may be possible, but for which data are not yet available. Lastly, distribution transformers do show significant savings because of the scale of distribution losses at the national level, even though total efficiency improvement is small on a percentage basis, due in part to the recent standards rulemaking.

4.2. Consumer Financial Impacts

By construction, the *Business Case* implements energy efficiency in a way that is cost-effective to consumers. Because this study insisted on quantifying investments needed to improve efficiency relative to the base case technology, the necessary information to evaluate these investments and financial benefits of energy savings, and therefore net financial impacts to consumers, is available for all appliance groups considered.

Recalling the definition of cost of conserved energy from Equation 1:

$$CCE = \frac{I \times q}{S}$$

The denominator of this equation $I \times q$ is the annualized equipment investment necessary to yield an annual energy savings S. BUENAS calculates the total savings $S_T(y)$ in each year, given by:

$$S_T(y) = S \times Stock'(y)$$

In this equation, *Stock'(y)* is the affected stock, that is, the number of units operating in the stock that were installed after the policy implementation date, and are each providing a savings S relative to the *Base Case*. Likewise, the total annualized investment in each year $I_T(y) \times q$ is given by:

$$I_T(y) \times q = I \times q \times Stock'(y)$$

Substituting Equation 1, and cancelling terms, yields:

$$I_T(y) \times q = S_T(y) \times CCE$$

In other words, total annualized investment can be calculated for each appliance group by multiplying its total energy savings by the cost of conserved energy shown in Table 5.

Financial savings from energy savings is given simply by the utility price in each year multiplied by the total energy savings $S_T(y)$. Net financial impacts are then given by:

$$N(y)=S_T(y) \times (Utility\ Price - CCE)$$

Costs, Savings and Net Impacts calculated in this way are shown in Table 5. In evaluating the financial value of efficiency or other government programs, it is customary to take account of deferred benefits through a discount rate calculation. The resulting *Net Present Value* (NPV) of benefits is given by:

$$NPV = \sum_{y=2010}^{2030} \frac{N(y)}{(1+DR)^{y-2010}}$$

In this equation, *DR* is a "societal" discount rate that parameterizes the preference for immediate returns on public investments. We consider two scenarios in which the societal discount rate is taken to be 3% or 7%. Cumulative equipment costs, energy bill savings, net savings and *NPV* are shown in Table 5. The table shows positive net savings for all appliance groups, which is not surprising, since the target efficiency levels were constructed to be cost-effective. The table shows that cost-effective efficiency improvements require an investment of 259 billion dollars over the next 20 years, but these investments will return over twice as much over the same period, for a net savings of 304 billion dollars, or of order of a thousand dollars per capita. The net present value is 214 billion dollars assuming a discount rate of 3%, and 140 billion with a 7% discount rate. Of the appliance groups studied, electric water heaters require the largest investment at $71.4 billion, but also have the highest payoff at $97.9 billion. Phasing out incandescent lamps has a similar payoff of $95.1 billion, but requires a`n investment of only $15.9 billion. This technology has the highest net benefit, of $79.2 billion, with commercial lighting providing the second highest, with $55.9 billion.

Finally, financial impacts, emissions savings and their relationship can be shown using a "conservation supply curve". This unique way of expressing the cost and benefits of carbon mitigation measures has become very widespread in the literature because of the key information it conveys. A conservation supply curve for the *Business Case* is presented in Figure 6. The x- axis shows cumulative carbon mitigation and expresses the relative importance of each appliance group. The total extent of the curve is 3021 mt CO_2, as shown in Table 4. The y-axis displays relative affordability according to cost of conserved energy. The blocks corresponding to each measure are ordered with increasing cost of conserved energy, from left to right. For comparability, cost of conserved energy for all end uses is expressed as dollars per kWh. The conversion factor for natural gas is 3412 Btu per kWh.

Table 5. Cumulative Financial Impacts of Efficiency Improvement through 2030

Appliance Group	Cumulative Financial Impacts				
	Cost	Savings	Net	NPV @3% DR	NPV @ 7% DR
	$ Billions				
Incandescent Lamps	15.9	95.1	79.2	66.4	53.2
Commercial Lighting	9.4	67.9	58.4	37.9	22.1
Standby Power	9.3	51.2	41.9	27.5	16.3
Distribution Transformers	41.4	76.4	34.9	22.6	13.1
Televisions	6.0	33.1	27.1	18.4	11.3
Electric Water Heater	71.4	97.9	26.5	17.2	10.1
Central AC & HP	35.0	50.5	15.4	10.0	5.9
Refrigerators	31.3	38.4	7.1	4.6	2.7
Room AC	6.1	11.0	4.9	3.2	1.9
Gas Water Heaters	8.9	12.1	3.2	2.1	1.2
Furnaces	20.0	23.0	3.0	2.0	1.1
Commercial Boilers	1.0	2.7	1.7	1.1	0.6
Commercial AC & HP	3.5	4.5	1.0	0.7	0.4
Electric Cooking Equipment	0.1	0.1	0.1	0.04	0.02
Total	**259**	**564**	**304**	**214**	**140**

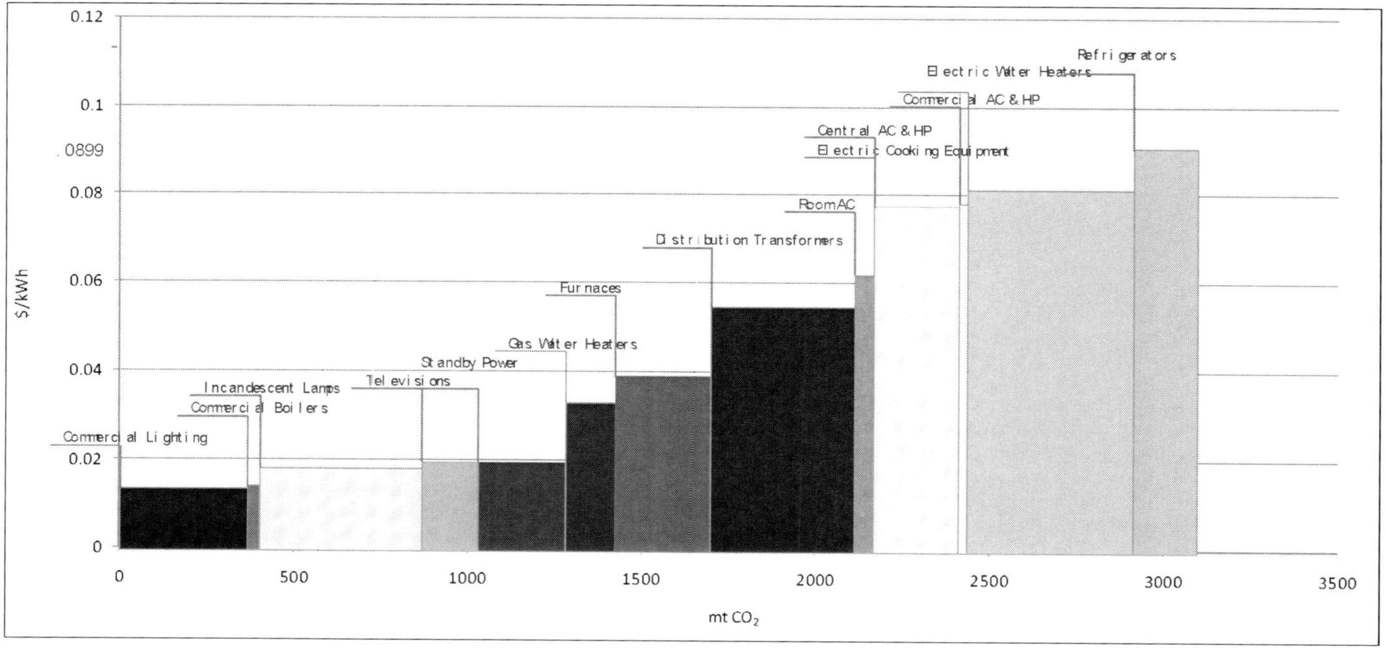

Figure 6. Cost of Conserved Energy for All End Uses 2010-2030.

Finally, we note that there are other benefits to the energy savings achieved in the *Business Case* besides the direct energy and financial benefits. The effect of reduction of greenhouse gas emissions and resulting avoided costs are difficult to quantify, but could be very large. One metric to consider the order of magnitude of the value of these types of impacts is the assumption of a carbon price. The assumption of a price of 25 dollars per ton of carbon dioxide yields an additional $76 billion dollars of savings, while a 100 dollar per ton price yields $302 billion additional dollars, doubling the total.

The negative impacts of emissions of SO_2 and NO from power plants are well-known (see, for example, EPA 2011), including acid rain, acidification of watersheds and lakes, and respiratory illness from inhaling particulates. Likewise, the reduction of mercury emissions from coal- burning power plants reduces fish contamination, which is now recognized as a major health risk. We do not try to quantify the health impacts of reduction of these emissions, only point out the obvious – that savings due to efficiency is equivalent to installation of clean electricity generation. In the *Business Case*, this reduction provides a large net financial benefit to consumers, which may not be true with alternatives.

Lastly, we consider the impacts of efficiency-related financial savings on the U.S. economy, and particularly in terms of job creation. In general, it is often stated that consumer spending generates growth and creates jobs at a higher rate than other types of expenditures. When efficiency improvements are implemented at the level of equipment purchased by residences and businesses, the result is a net savings distributed over a very large number of consumers. Presumably, much of this savings, which comes in the form of lowered utility bills, will be injected back into the economy in the form of purchased consumer goods. A recent study found that:

> "The principle energy-related sectors of the U.S. economy are not especially job-intensive in comparison to the rest of the economy. They support only 7.4 total jobs for every one million dollars of revenue received in utility bill payments in comparison with the rest of the economy, which supports between 13 and 21 direct jobs per million dollars of receipts". (Gold 2011 p.8)

Using this result as a guide, we deduce that energy savings can be translated into job creation at the rate of 5.6-13.6 jobs per million dollars. A net savings of 304 billion dollars over 20 years corresponds to 15.2 billion dollars per year which supports 85,000 to 207,000 jobs for the entire period.

5. CONCLUSIONS

The *Business Case* analysis found additional potential for cost-effective efficiency improvement in the United States for fourteen appliance groups in the residential and commercial building sectors. Efficiency improvement for these technologies could deliver twice as much financial benefit to U.S. households and business than the investment needed to implement them. In addition to direct financial benefits, impacts on greenhouse gas emissions and job creation are significant. Total net impacts from additional deployment of high efficiency technology include:

Final Energy savings:

- 230 billion kWh of electricity and 0.14 EJ of natural gas per year in 2020
- 430 billion kWh of electricity and 0.35 EJ of natural gas per year in 2030
- A total of 4900 billion kWh of electricity and 3.1 EJ natural gas cumulatively through 2030

Primary Energy Savings:

- 1.3 EJ of coal per year in 2020 and 0.58 EJ of natural gas per year in 2020
- 2.4 EJ of coal per year in 2030 and 1.2 EJ of natural gas per year in 2030
- A total of 27 EJ of coal and 13 EJ of natural gas cumulatively through 2030

Cumulative greenhouse gas emissions mitigation:

- 3000 million metric tons of CO_2 through 2030
- 12 million metric tons of SO_2 through 2030

- 3.5 million metric tons of NO through 2030
- 62 metric tons of mercury through 2030

Financial impacts to consumers through 2030:

- Equipment investment of 260 billion dollars
- Energy bill savings of 560 billion dollars
- Net savings of 300 billion dollars

Job Creation:

- Net creation of 85,000 to 200,000 jobs

The end uses studied in this report are concentrated in the buildings sector, and the bulk of the savings is to be found in electricity savings. The study finds that by 2030, cost-effective efficiency improvements could save 430 billion TWh, which represents about 12% of total buildings electricity consumption in that year, according to the International Energy Agency's World Energy Outlook[30]. This corresponds to 9.3% of electricity consumption in the U.S. or 3.7% of total energy consumption, in primary energy terms. In the same year, the *Business Case* saves 265 mt CO_2 in 2030. According to WEO, with current policies in place, emissions in the United States will total 5310 mt CO_2 or 712 mt more than the Annex I target of 4598 mt (5.2% below the 1990 level of 4850 mt. Implementation of the *Business Case* would close over a third of this gap. In addition to the 300 billion dollars net savings provided to consumers through 2030, annual savings of 430 billion kWh would avoid capital investments of 32 billion dollars[31].

Since most of the equipment studied has been the subject of at least one efficiency standard, this finding supports the notion of efficiency improvement as an ongoing process. For example, although residential refrigerators have improved dramatically in efficiency over the last 30 years, there is room for improvement.

It should also be noted that many of the technologies included in the *Business Case* scenario were not available ten to twenty years ago, or at least weren't shown to be cost-effective. These technologies have become available and cost-effective through research, new materials and components, improvements in production processes, or changes in design of systems. We expect that a similar analysis performed 10 years from now will show improvements not accessible to the current study due either to lack of data or prohibitively high cost of "prototype" technologies.

Two main caveats must be made in the interpretation of these results. First, because the rigor of the methodology used to evaluate cost-effectiveness requires a significant amount of technical data, we only cover a subset of equipment types for which significant savings potential might be available. In particular, the appliance groups covered are limited to buildings applications. For this sector, however, we believe a large fraction of energy demand is accounted for. For this reason, while the overall savings potential is large, it cannot be interpreted as "comprehensive". Second, most of the data used are sourced from DOE rulemakings to set mandatory energy performance standards. As a consequence, only those product types already subject to standards are covered, introducing a bias towards equipment with a relatively high efficiency baseline. Furthermore, much of the data are for products with a recent standard, or one that is expected in the near future, limiting the amount of savings that can be considered additional. In some cases, the cost-effective efficiency level may exceed the standards set by DOE. We emphasize that this is not to be taken as a claim that DOE standards are less stringent than what is economically justified, rather that the simplified analysis performed in the current study is representative of the potential additional savings, but cannot be precisely compared. In some cases, technologies found by DOE to be attractive for standards on a purely cost-benefit basis were not adopted as standards due to other barriers, such as the absence of wide-scale production capability by manufacturers. The current analysis does not investigate these, but implicitly assumes that many of these barriers are removed by 2015.

Finally, we believe this study to be among the few to attempt to evaluate the "economic" potential of efficiency improvement on a national scale and in a transparent way. In addition to demonstrating significant savings potential, we hope that it demonstrates a clear and consistent methodology for creation and expansion of alternative energy scenarios in the U.S. and beyond. Additional scenarios that could be explored include the potential impact of carbon taxes, cap- and-trade, R&D investments and other policy- or market-based drivers. The ability of the research community to utilize this type of analysis to inform government and private sector decision makers will depend largely on investments made in development of the type of data used here, both more widely and with greater frequency.

APPENDIX 1. U.S. DEPARTMENT OF ENERGY APPLIANCE STANDARDS SCHEDULE

Product Covered	Initial Legislation	Last MEPS issued	Effective Date	Update Expected	Potential Effective Date
Residential Products					
Battery Chargers	EPACT 2005	None	None	2011	2013
Boilers	NAECA 1987	2007	2012	2015	2020
Central AC and Heat Pumps	NAECA 1987	2001	2006	2011	2015
Clothes Dryers	NAECA 1987	1991	1994	2011	2014
Clothes Washers	NAECA 1987	2007	2011	2011	2015
	EPACT 2005				
Compact Audio Equipment	None				
Dehumidifiers	EPACT 2005	2007	2012	2015	2018
Direct Heating Equipment	NAECA 1987	2010	2013	2018	2023
Dishwashers	NAECA 1987	2007	2010	2015	2018
DVD Players and Recorders	None				
External Power Supplies	EPACT 2005	2007	2008	2011	2013
Furnaces	NAECA 1987	2007	2015	2011	2013
Furnace Fans	EPACT 2005	None	None	2013	2016
Microwave Ovens	NAECA 1987	None	None	2011	2014
Plumbing Products	EPACT 1992	1992	1994		
Pool Heaters	NAECA 1987	2010	2013	2018	2021
Pool Pumps	None				
Portable Electric Spas (Hot Tubs)	None				
Ranges and Ovens (electric)	NAECA 1987	None	None	2017	2020
Ranges and Ovens (gas)	NAECA 1987	2009	2012	2014	2017
Refrigerators & Freezers	NAECA 1987	1997	2001	2010	2014
Room AC	NAECA 1987	1997	2000	2011	2014
Televisions	None			2013	2016
Water Heaters	NAECA 1987	2010	2015	2018	2023
Commercial Products					
Automatic Commercial Ice Makers	EPACT 2005	2005	2010	2010	2015
Commercial 3-Phase Central AC	EPACT 1992	2007	2008	2015	2018

Product Covered	Initial Legislation	Last MEPS issued	Effective Date	Update Expected	Potential Effective Date
Commercial AC & Heating	EPACT 1992	2005	2010	2013	2016
Equipment (Air-cooled)					
Commercial AC & Heating Equipment (Water- and evaporatively-cooled)	EPACT 1992	2009	2011	2017	2020
Commercial Boilers	EPACT 1992	2009	2012	2017	2020
Commercial Clothes Washers	EPACT 2005	2010	2013	2015	2018
Commercial Warm Air Furnaces Commercial Refrigeration	EPACT 1992	2001	2003	2009	2012
Supermarket (built-up) Refrigerators & Freezers	EPACT 2005	2009	2012	2013	2016
Refrigerator and Freezer Cases without Doors	EPACT 2005	2009	2012	2013	2016
Reach-In Refrigerators & Freezers	EPACT 2005	2005	2010	2013	2016
Walk-In Refrigerators Freezers	EISA 2007	2007	2009	2012	2015
Commercial Water Heaters Distribution Transformers	EPACT 1992	2001	2003	2009	2012
Low-V, Dry-Type	EPACT 2005	2005	2007	2013	2016
Medium-V, Dry-type	EPACT 1992	2007	2010	2012	2015
Liquid Immersed	EPACT 1992	2007	2010	2012	2015
Electric Motors (> 1 hp)	EPACT 1992	2007	2010	2012	2015
Electric Motors, small (<= 1 hp)	EPACT 1992	2010	2015	2018	2021
Hot Food Holding Cabinets	None				
Packaged Terminal AC and Heat Pumps	EPACT 1992	2008	2010/201	2016	2019
Pre-Rinse Spray Valves	EPACT 2005	2005	2006	2013	2016
Unit Heaters	EPACT 2005	2005	2008	2013	2016
Vending Machines	EPACT 2005	2009	2012	2017	2020
Water Dispensers	None				
Lighting Products					
65 watt BR and ER IRL's and IRL's < 50 watts	None	None	None	2010	2013

(Continued)

Product Covered	Initial Legislation	Last MEPS issued	Effective Date	Update Expected	Potential Effective Date
Ceiling Fans and Ceiling Fan Light Kits	EPACT 2005	2005	2007	2013	2016
Compact Fluorescent Lamps	EPACT 2005	2005	2006	2013	2016
Fluorescent Lamp Ballasts	NAECA 1987	2000	2005	2011	2014
	EPACT 2005				
General Service Incandescent Lamps plus CFL, GSLED, GSOLED	EPACT 1992	2007	2012	2017	2020
HID Lamps	EPACT 1992	None	None	2014	2017
Illuminated Exit Signs	EPACT 2005	2005	2006	2013	2016
Incandescent Reflector Lamps (IRL)	EPACT 1992	2009	2012	2014	2017
Linear Fluorescent Lamps	EPACT 1992	2009	2012	2014	2017
Mercury Vapor Lamp Ballasts	EPACT 2005	2005	2008	N/A	N/A
Metal Halide Lamp Fixtures	EISA 2007	2007	2009	2012	2015
Torchiere Lamps	EPACT 2005	2005	2006	2013	2016
Traffic Signals	EPACT 2005	2005	2006	2013	2016

Source: ASAP (2011).

APPENDIX 2. EFFICIENCY-COST RELATIONSHIP AND COST OF CONSERVED ENERGY CALCULATION FOR APPLIANCE GROUPS

Parameters used in Calculation of Cost of Conserved Energy:

Residential Consumer Discount Rate = 5%
Commercial Consumer Discount Rate = 6%
Residential Electricity Price (2015) = $0.11 $/kWh
Commercial Electricity Price (2015) = $0.10 $/kWh
Residential Natural Gas Price (2015) $13.17 $/MMBtu
Commercial Natural Gas Price (2015) $11.40 $/MMBtu

Table A.2.1. Efficiency-Cost Relationship for Residential Refrigerators

Class	**Top Mount Auto Defrost Refrigerator - Freezer**			**Bottom-Mount Refrigerator-Freezer**			**Side-by-Side Refrigerator-Freezer**		
Market share	**54.3%**			**13.5%**			**28.9%**		
Lifetime	**17.1**			**17.1**			**17.1**		
q	0.088			0.088			0.088		
Efficiency Level	**UEC kWh/yr**	**Price $2010**	**CCE $2010**	**UEC kWh/yr**	**Price $2010**	**CCE $2010**	**UEC kWh/yr**	**Price $2010**	**CCE $2010**
Baseline	501	547		613	953		768	1161	
1	451	559	0.021	551	955	0.003	691	1164	0.003
2	426	567	0.024	521	957	0.004	653	1169	0.006
3	401	629	0.072	490	963	0.007	614	1188	0.015
4	379	672	0.091	459	1028	0.043	576	1254	0.042
5	351	765	0.128	429	1136	0.087	538	1396	0.090
6	323	899	0.175	391	1286	0.132	515	1508	0.120
Target Annual Incremental Cost (I×q)	$11			$16			$21		
Target UEC (kWh/yr)	379			429			538		

Weighted Average Baseline UEC	588 kWh/yr
Weighted Average Target UEC	428 kWh/yr
Weighted Average Energy Savings	161 kWh/yr
Weighted Average Incremental Cost	14.5 $2010
Product CCE	0.090 $2010

Table A.2.2. Efficiency-Cost Relationship for Residential Room Air Conditioners

Class	**< 6000 Btu/Hr with Louvered sides**			**8000-13999 Btu/Hr with Louvered sides**			**≥20,000 Btu/Hr, with Louvered sides**			**8000-13999 Btu/Hr without Louvered sides**		
Market share	41.0%			44.6%			3.8%			10.6%		
Lifetime	10.5			10.5			10.5			10.5		
q	0.125			0.125			0.125			0.125		
Efficiency Level	**UEC**	**Price**	**CCE**	**UEC**	**Price**	**CCE**	**UEC**	**Price**	**CCE**	**UEC**	**Price**	**CCE**
	kWh/yr	**$2010**	**$2010**	**kWh/yr**	**$2010**	**$2010**	**kWh/yr**	**$2010**	**$2010**	**kWh/yr**	**$2010**	**$2010**
Baseline	**688**	**335**		**672**	**497**		**1118**	**875**		**712**	**567**	
1	648	343	0.025	632	512	0.044	1052	899	0.046	673	572	0.016
2	618	353	0.033	602	526	0.051	1007	929	0.061	644	579	0.022
3	590	368	0.043	576	552	0.071	966	967	0.076	611	600	0.041
4	565	386	0.052	566	612	0.135	947	1013	0.101			
5	546	396	0.054									
Target Annual Incremental Cost (I×q)	$8			$7			$17			$4		
Target UEC (kWh/yr)	546			576			947			611		

Weighted Average Baseline UEC	700 kWh/yr
Weighted Average Target UEC	581 kWh/yr
Weighted Average Energy Savings	118 kWh/yr
Weighted Average Incremental Cost	7.27 $2010
Product CCE	0.062 $2010

Table A.2.3. Efficiency-Cost Relationship for Residential Water Heaters

Class	Electric Water Heater		
Market share	**100.0%**		
Lifetime	**13**		
q	**0.106**		
Efficiency Level	**UEC**	**Price**	**CCE**
	kWh/yr	**$2010**	**$2010**
Baseline	2604	666	
1	2569	677	0.033
2	2535	688	0.034
3	2515	695	0.035
4	2467	730	0.049
5	2431	763	0.060
6	1399	1589	0.081
7	1216	1717	0.081
8	1216	1717	0.081
Target Annual Incremental Cost (I×q)	$112		
Target UEC (kWh/yr)	1216		

Weighted Average Baseline UEC	2604 kWh/yr
Weighted Average Target UEC	1216 kWh/yr
Weighted Average Energy Savings	1388 kWh/yr
Weighted Average Incremental Cost	112 $2010
Product CCE	0.081 $2010

Class	Gas Water Heater			Gas-Fired Instantaneous Water Heater		
Market share	90.2%			9.1%		
Lifetime	13			13		
q	0.106			0.106		
Efficiency Level	**UEC**	**Price**	**CCE**	**UEC**	**Price**	**CCE**
	MMBtu	**$2010**	**$2010**	**MMBtu**	**$2010**	**$2010**
Baseline	16.5	1185		16.7	2605	
1	15.8	1245	8.69	14.8	2605	0.00
2	15.3	1302	10.55	11.5	2616	0.23
3	14.9	1580	25.54	11.2	2626	0.41
4	14.5	1612	22.11	10.9	2633	0.52
5	13.9	1672	20.15	10.7	2986	6.70
6	12.4	1908	18.53	10.6	3050	7.67
7	15.7	1263	9.73	9.8	2943	5.16
8	15.2	1320	11.18	9.4	3122	7.55
Target Annual Incremental Cost (I×q)	$14			$55		
Target UEC (MMBtu/yr)	15			9		

Weighted Average Baseline UEC	16.6 MMBtu/yr
Weighted Average Target UEC	14.7 MMBtu/yr
Weighted Average Energy Savings	1.87 MMBtu/yr
Weighted Average Incremental Cost	8.1 $2010
Product CCE	9.68 $2010

Table A.2.4. Efficiency-Cost Relationship for Residential Electric Cooking Equipment

Class	Electric Coil Cooktops			Electric Smooth Cooktops			Electric Standard Ovens			Electric Self-Cleaning Ovens			Microwave Ovens		
Market share	9.7%			11.4%			5.6%			17.3%			56.0%		
Lifetime	19			19			19			19			9		
q	0.083			0.083			0.083			0.083			0.141		
Efficiency Level	**UEC**	**Price**	**CCE**	**UEC**	**Price**	**CCE**	**UEC**	**Price**	**CCE**	**UEC**	**Price**	**CCE**	**UEC**	**Price**	**CCE**
	kWh /yr	**$2010**	**$2010**	**kWh /yr**	**$2010**	**$2010**	**kWh /yr**	**$2010**	**$2010**	**kWh /yr**	**$2010**	**$2010**	**kWh /yr**	**$2010**	**$2010**
Baseline	**128**	**292**		**128**	**331**		**167**	**443**		**171**	**362**		**156**	**235**	
1	123	295	0.055	126	589	11.225	160	445	0.033	171	369	1.403	149	249	0.293
2							154	450	0.049	168	431	1.822	149	263	0.569
3							152	456	0.075				147	286	0.811
4							149	517	0.354				146	315	1.142
5							149	523	0.371						
Target Annual Incremental Cost (I×q)	$0.29			$0.00			$1.10			$0.00			$0.00		
Target UEC (kWh/yr)	123			No option with CCE below utility price			152			No option with CCE below utility price			No option with CCE below utility price		

Weighted Average Baseline UEC	153 kWh/yr
Weighted Average Target UEC	152 kWh/yr
Weighted Average Energy Savings	1.33 kWh/yr
Weighted Average Incremental Cost	0.090 $2010
Product CCE	0.067 $2010

Table A.2.5. Efficiency-Cost Relationship for Residential Central Air Conditioners and Heat Pumps

Class	Split System AC (Coil Only)			Split System AC (Blower-Coil)			Single Package AC			Split System HP			Single Package HP		
Market share	56.7%			6.3%			5.9%			27.6%			3.5%		
Lifetime	19.01			19.01			19.01			19.01			19.01		
q	0.083			0.083			0.083			0.083			0.083		
Efficiency Level	UEC	Price	CCE	UEC	Price	CCE	UEC	Price	CCE	UEC	Price	CCE	UEC	Price	CCE
	kWh /yr	$2010	$2010	kWh /yr	$2010	$2010	kWh /yr	$2010	$2010	kWh /yr	$2010	$2010	kWh /yr	$2010	$2010
Baseline	2407	2799		2433	4126		2376	2649		5236	3296		5184	3009	
1	2322	2873	0.071	2347	4201	0.072	2292	2710	0.060	5079	3381	0.045	5029	3090	0.043
2	2243	2949	0.075	2267	4276	0.075	2214	2774	0.064	4933	3471	0.048	4884	3195	0.051
3	2168	3038	0.083	2190	4357	0.079	2139	2859	0.073	4792	3540	0.045	4774	3345	0.068
4	2097	3139	0.091	2119	4416	0.076	2070	2983	0.090	4660	3606	0.045	4672	3477	0.076
5	2031	3262	0.102	2052	4477	0.076	2004	3085	0.097	4536	3678	0.045	4481	3630	0.073
6	1968	3419	0.117	1988	4541	0.077	1942	3212	0.107	4418	3772	0.048	4305	3811	0.076
7	1912	3599	0.134	1932	4606	0.079	1876	3362	0.118	4173	3930	0.049	4239	3993	0.086
8				1878	4673	0.082				3954	4100	0.052			
9				1780	4816	0.087				3758	4282	0.055			
10				1692	4969	0.094				3673	4381	0.057			
11				1613	5133	0.102									
12				1541	5309	0.110									
13				1476	5498	0.119									
Target Annual Incremental Cost (I×q)	$38			$98			$47			$90			$81		
Target UEC (kWh/yr)	2031			1541			1942			3673			4239		

Weighted Average Baseline UEC 3285 kWh/yr

Weighted Average Target UEC	2525 kWh/yr
Weighted Average Energy Savings	759 kWh/yr
Weighted Average Incremental Cost	58.2 $2010
Product CCE	0.077 $2010

Table A.2.6. Efficiency-Cost Relationship for Residential Gas Furnaces

Class	Non-Weatherized Gas Furnace			Weatherized Gas Furnace			Mobile home Gas Furnace			Oil-fired Furnace		
Market share	84.0%			9.5%			3.6%			3.0%		
Lifetime	20			18			19			15		
q	0.080			0.086			0.083			0.096		
Efficiency Level	UEC MMBtu	Price $2010	CCE $2010	UEC MMBtu	Price $2010	CCE $2010	UEC MMBtu	Price $2010	CCE $2010	UEC MMBtu	Price $2010	CCE $2010
Baseline	53.9	2530		33.7	4169		40.9	1010		73.2	3439	
1	53.9	2530	0.72	33.5	4170	0.98	40.1	1025	1.57	73.1	3439	0.26
2	53.9	2735	-233.41	33.2	4180	1.89	40.2	1405	46.57	72.6	3444	0.88
3	53.5	2581	10.21	32.8	4301	12.49	39.6	1119	7.16	72.7	3462	4.41
4	53.6	2788	65.40	32.4	4342	11.52	39.7	1505	35.23	72.8	3681	59.45
5	50.4	3003	10.84				39.2	1150	6.80	72.0	3447	0.69
6	49.7	3141	11.65				39.3	1535	26.91	72.1	3477	3.37
7	49.7	3416	17.07				36.0	1397	6.56	72.2	3710	26.53
8	49.6	3462	17.66							71.3	3595	8.17
9	47.7	4167	21.37							71.4	3628	10.44
10										71.5	3883	26.26
11										70.6	3756	12.04
12										70.7	3791	13.94
13										70.8	4069	26.16

Table A.2.6. (Continued)

Efficiency Level	UEC	Price	CCE	UEC	Price	CCE	UEC	Price	CCE	UEC	Price	CCE
	MMBtu	$2010	$2010	MMBtu	$2010	$2010	MMBtu	$2010	$2010	MMBtu	$2010	$2010
14										69.9	3939	14.72
15										70.0	3977	16.35
16										70.1	4277	26.40
Target Annual Incremental Cost (I×q)	$49			$15			$32			$31		
Target UEC (MMBtu/yr)	49.7			32.4			36.0			70.6		

Weighted Average Baseline UEC	52.1 MMBtu/yr
Weighted Average Target UEC	48.2 MMBtu/yr
Weighted Average Energy Savings	3.91 MMBtu/yr
Weighted Average Incremental Cost	44.6 $2010
Product CCE	11.42 $2010

Table A.2.7. Efficiency-Cost Relationship for Commercial Building Air Conditioners and Heat Pumps

Class	≥65,000 and <135,000 Btu/Hr Unitary Air Conditioners			≥135,000 and <240,000 Btu/Hr Unitary Air Conditioners			Packaged Terminal Air Conditioners Standard, 9 kBtu/Hr			Packaged Terminal Air Conditioners Standard, 12 kBtu/Hr		
Market share	30.5%			13.3%			13.8%			20.3%		
Lifetime	15.4			15.4			10			10		
q	0.101			0.101			0.136			0.136		
Efficiency Level	UEC	Price	CCE	UEC	Price	CCE	UEC	Price	CCE	UEC	Price	CCE
	kWh /yr	$2010	$2010	kWh /yr	$2010	$2010	kWh /yr	$2010	$2010	kWh /yr	$ 2010	$2010
Baseline	13857	9046		29854	14774		1013	1235		1294	1476	
1				28733	15426	0.016	1001	1244	0.097	1279	1488	0.113

Class	≥65,000 and <135,000 Btu/Hr Unitary Air Conditioners			≥135,000 and <240,000 Btu/Hr Unitary Air Conditioners			Packaged Terminal Air Conditioners Standard, 9 kBtu/Hr			Packaged Terminal Air Conditioners Standard, 12 kBtu/Hr		
2				27714	16420	0.020	990	1256	0.102	1264	1501	0.104
3										1251	1515	0.125
Target Annual Incremental Cost (I×q)	$0			$167			$1.84			$2.26		
Target UEC (kWh/yr)	No option with CCE below utility price			27714			No option with CCE below utility price			No option with CCE below utility price		

Class	Packaged Terminal Air Conditioners Non-standard, 11 kBtu/Hr			Packaged Terminal Heat Pumps Standard, 9 kBtu/Hr		
Market share	2.1%			10.8%		
Lifetime	10			10		
q	0.136			0.136		
Efficiency Level	**UEC**	**Price**	**CCE**	**UEC**	**Price**	**CCE**
	kWh/yr	**$2010**	**$2010**	**kWh/yr**	**$2010**	**$2010**
Baseline	1658	1575		1984	1358	
1	1584	1599	0.044	1957	1367	0.047
2	1559	1611	0.050	1945	1377	0.069
3	1536	1626	0.057			
Target Annual Incremental Cost (I×q)	$6.92			$2.68		
Target UEC (kWh/yr)	1536			1945		

Weighted Average Baseline UEC	8901 kWh/yr
Weighted Average Target UEC	8608 kWh/yr
Weighted Average Energy Savings	293 kWh/yr
Weighted Average Incremental Cost	22.7 $2010
Product CCE	0.078 $2010

Table A.2.8. Efficiency-Cost Relationship for Commercial Building Fluorescent Lamp Ballasts

Class	Two-Lamp RS and IS Ballasts Operating 4-Foot MBP and 2-Foot U-Shaped Lamps in the Commercial Sector (F32T8 Baseline)			Two-Lamp PS Ballasts Operating 4- Foot MBP and 2-Foot U-Shaped Lamps in the Commercial Sector			Four-Lamp RS and IS Ballasts Operating 4-Foot MBP and 2-Foot U-Shaped Lamps in the Commercial Sector			Four-Lamp PS Ballasts Operating 4-Foot MBP and 2- Foot U-Shaped Lamps in the Commercial Sector		
Market share	54.3%			6.4%			24.0%			3.5%		
Lifetime	15			15			15			15		
q	0.103			0.103			0.103			0.103		
Efficiency Level	**UEC**	**Price**	**CCE**	**UEC**	**Price**	**CCE**	**UEC**	**Price**	**CCE**	**UEC**	**Price**	**CCE**
	kWh /yr	**$ 2010**	**$ 2010**	**kWh /yr**	**$ 2010**	**$ 2010**	**kWh /yr**	**$2010**	**$ 2010**	**kWh /yr**	**$ 2010**	**$ 2010**
Baseline	208	57		153	60		383	103		289	107	
1	196	62	0.048	142	63	0.027	380	109	0.205	275	109	0.019
2	194	60	0.025									
Target Annual Incremental Cost (I×q)	$0.34			$0.28			-$1.74			$0.28		
Target UEC (kWh/yr)	194			142			No option with CCE below utility price			275		

Class	Two-Lamp PS Ballasts Operating 4-Foot T5 MiniBP SO Lamps in the Commercial Sector			Two-Lamp PS Ballasts Operating 4- Foot T5 MiniBP HO Lamps in the Industrial Sector			Four-Lamp RS and IS Ballasts Operating 8-Foot RDC HO Lamps (Cold Temperature/ Sign Ballasts) in the Commercial Sector		
Market share	3.3%			3.3%			2.9%		
Lifetime	15			10			0.15		
q	0.103			0.136			0.103		
Efficiency Level	**UEC**	**Price**	**CCE**	**UEC**	**Price**	**CCE**	**UEC**	**Price**	**CCE**
	kWh/yr	**$2010**	**$2010**	**kWh/yr**	**$2010**	**$2010**	**kWh/yr**	**$2010**	**$2010**
Baseline	267	68		688	71		1352	228	
1	223	71	0.005	598	73	0.004	1113	218	-0.005
2	215	82	0.027	571	81	0.012			
Target Annual Incremental Cost (I×q)	$0.28			$0.42			-$1.08		
Target UEC (kWh/yr)	215			571			1113		

Weighted Average Baseline UEC	306 kWh/yr
Weighted Average Target UEC	285 kWh/yr
Weighted Average Energy Savings	21.2 kWh/yr
Weighted Average Incremental Cost	0.212 $2010
Product CCE	0.010 $2010

Table A.2.9. Efficiency-Cost Relationship for High-Intensity Discharge Lamps

Class	Low Wattage 1			Low Wattage 2			Low Wattage 3			Small Area			Roadway Arch		
Market share	3.7%			3.0%			2.9%			5.9%			6.1%		
Lifetime	13.6			9.8			12.1			12.1			12.1		
q	0.110			0.138			0.119			0.119			0.119		
Efficiency Level	**UEC**	**Price**	**CCE**	**UEC**	**Price**	**CCE**	**UEC**	**Price**	**CCE**	**UEC**	**Price**	**CCE**	**UEC**	**Price**	**CCE**
	kWh /yr	**$2010**	**$2010**	**MMbtu**	**$2010**	**$2010**	**MMbtu**	**$2010**	**$2010**	**MMbtu**	**$2010**	**$2010**	**MMbtu**	**$2010**	**$2010**
Baseline	1672	25		768	16		1875	25		861	16		861	16	
1	1413	66	0.017	749	88	0.537	1586	66	0.017	840	88	0.412	840	88	0.412
2	1672	76	0.000	694	28	0.024	1875	76		778	28	0.018	778	28	0.018
3	1303	133	0.032	461	39	0.011	1461	133	0.031	518	39	0.008	518	39	0.008
4	1542	84	0.049	697	23	0.015	1731	84	0.048	782	23	0.011	782	23	0.011
5	1524	36	0.008	461	39	0.011	1710	36	0.008	518	39	0.008	518	39	0.008
6	1354	43	0.006	343	20	0.001	1519	43	0.006	385	20	0.001	385	20	0.001
7	1092	20	-0.001	413	35	0.007	1225	20	-0.001						
8	1081	21	-0.001	321	9	-0.002	1213	21	-0.001						
9	1063	41	0.003				1192	41	0.003						
10	886	19	-0.001				994	19	-0.001						
11	1007	69	0.007												
12	878	40	0.002												
Target Annual Incremental Cost (I×q)	$1.64			-$0.95			-$0.71			$0.55			$0.55		
Target UEC (kWh/yr)	878			321			994			385			385		

Class	Street			Security			Security - Res			Landscape		
Market share	17.5%			16.6%			15.5%			7.0%		
Lifetime	12.1			12.1			12.1			12.1		
q	0.119			0.119			0.119			0.119		
Efficiency Level	**UEC**	**Price**	**CCE**	**UEC**	**Price**	**CCE**	**UEC**	**Price**	**CCE**	**UEC**	**Price**	**CCE**
	MMbtu	**$2010**	**$2010**	**MMbtu**	**$2010**	**$2010**	**MMbtu**	**$2010**	**$2010**	**MMbtu**	**$2010**	**$2010**
Baseline	861	14		861	14		861.12	16.00		861.12	13.55	
1	840	78	0.367	840	76	0.357	385.02	20.25	0.001	840.42	75.83	0.357
2	778	25	0.016	778	25	0.016				778.32	24.58	0.016
3	518	35	0.007	518	34	0.007				517.50	34.16	0.007
4	782	21	0.010	782	20	0.010				782.46	20.14	0.010
5	385	18	0.001	385	18	0.001				517.50	34.16	0.007
6										385.02	17.55	0.001
7												
8												
9												
10												
11												
12												
Target Annual Incremental Cost (I×q)	$0.49			$0.47			$0.50			$0.47		
Target UEC (kWh/yr)	385			385			385			No option with CCE below utility price		

Weighted Average Baseline UEC 729 kWh/yr
Weighted Average Target UEC 335 kWh/yr
Weighted Average Energy Savings 394 kWh/yr
Weighted Average Incremental Cost 0.242 $2010
Product CCE 0.038 $2010

Table A.2.10. Efficiency-Cost Relationship for Distribution Transformers

Class	Design Line 1 (Liquidimmersed, 50kVA, singlephase, rectangular tank)			Design Line 2 (Liquidimmersed, 25kVA, singlephase, round tank)			Design Line 4 (Liquidimmersed, 150kVA, threephase)		
Market share	35.9%			59.3%			3.0%		
Lifetime	32			32			32		
q	0.071			0.071			0.071		
Efficiency Level	**UEC**	**Price**	**CCE**	**UEC**	**Price**	**CCE**	**UEC**	**Price**	**CCE**
	kWh/yr	**$2010**	**$2010**	**kWh/yr**	**$2010**	**$2010**	**kWh/yr**	**$2010**	**$2010**
Baseline	5691	3301	0.012	1407	2143	0.007	9306	7411	0.037
1	4829	3639	0.020	884	2359	0.017	8856	8117	0.054
2	2231	5394	0.037	817	3916	0.104	8398	8117	0.044
3	4583	3301	0.005	1407	2143	0.007	5520	10972	0.058
4	5403	3348	0.012	1548	2285	0.028	5520	12013	0.072
5	5403	3348	0.012	1656	2285	0.035	8363	7411	0.023
6	5691	3639	0.039	1691	2511	0.081	8398	7411	0.023
7							8856	7411.02	0.029
8							8398	7411	0.023
Target Annual Incremental Cost (I×q)	$149			$126			$253		
Target UEC (kWh/yr)	2231			817			5520		

Weighted Average Baseline UEC	5702 kWh/yr
Weighted Average Target UEC	2720 kWh/yr
Weighted Average Energy Savings	2982 kWh/yr
Weighted Average Incremental Cost	162 $2010
Product CCE	0.054 $2010

Table A.2.11. Efficiency-Cost Relationship for Commercial Building Boilers

Class	Packaged Boilers, Gas-Fired, HW 400-1500 kBtu/Hr			Packaged Boilers, Gas-Fired, HW 3000 kBtu/Hr			Packaged Boilers, Gas-Fired, Steam (no natural draft) 400-1500 kBtu/Hr			Packaged Boilers, Gas-Fired, Steam (no natural draft) 3000 kBtu/Hr			Packaged Boilers, Gas-Fired, Steam (natural draft) 400-1500 kBtu/Hr		
Market share	24.2%			3.9%			8.2%			7.1%			12.6%		
Lifetime	30			30			30			30			30		
q	0.073			0.073			0.073			0.073			0.073		
	UEC	**Price**	**CCE**	**UEC**	**Price**	**CCE**	**UEC**	**Price**	**CCE**	**UEC**	**Price**	**CCE**	**UEC**	**Price**	**CCE**
Efficiency Level	**MMbtu**	**$2010**	**$2010**	**MMbtu**	**$2010**	**$2010**	**MMbtu**	**$2010**	**$2010**	**MMbtu**	**$2010**	**$2010**	**MMbtu**	**$2010**	**$2010**
Baseline	1035	23513		3958	44665		1077	27564		4025	67175		1111	26303	
1	1029	24471	11.60	3934	47147	7.51	1072	28973	20.47	4005	69057	6.84	1101	28686	17.31
2	1016	25939	9.28	3897	47147	2.96	1061	30799	14.69	3968	70655	4.44	1090	29742	11.90
3	998	29344	11.45	3858	57049	9.00	1050	33976	17.25	3931	72411	4.05	1077	31798	11.74
4	979	34966	14.86	3671	86005	10.46	1038	36557	16.75	3893	74500	4.03			
Target Annual Incremental Cost (I×q)	$176			$3,003			$0			$532			$0		
Target UEC (MMBtu/yr)	1016			3671			No option with CCE below utility price			3893			No option with CCE below utility price		

Table A.2.11. (Continued)

Class	Packaged Boilers, Gas-Fired, Steam (natural draft) 3000 kBtu/Hr			Packaged Boilers, Oil-Fired, HW 400-1500 kBtu/Hr			Packaged Boilers, Oil-Fired, Steam 400-1500 kBtu/Hr			Packaged Boilers, Oil-Fired, Steam 3000 kBtu/Hr		
Market share	9.1%			6.8%			11.4%			15.0%		
Lifetime	30			30			30			30		
q	0.073			0.073			0.073			0.073		
Efficiency Level	**UEC MMbtu**	**Price $2010**	**CCE $2010**	**UEC MMbtu**	**Price $2010**	**CCE $2010**	**UEC MMbtu**	**Price $2010**	**CCE $2010**	**UEC MMbtu**	**Price $2010**	**CCE $2010**
Baseline	4111	69953		1022	25278		1052.00	24373.30		3920.00	67593.40	
1	4104	70451	5.26	1007	27898	12.69	1046.00	25964.21	19.26	3904.00	69226.02	7.41
2	4074	71974	4.04	990	30165	11.10	1037.00	27540.88	15.34	3875.00	71596.11	6.46
3	4035	75037	4.94	968	34793	12.80	1014.00	31800.94	14.20	3836.00	76772.68	7.94
4	3989	80239	6.23							3757.00	86603.98	8.47
Target Annual Incremental Cost (I×q)	$760			$355			$0			$1,381		
Target UEC (MMBtu/yr)	3989			990			No option with CCE below utility price			3757		

Weighted Average Baseline UEC	2106 MMBtu/yr
Weighted Average Target UEC	2042 MMBtu/yr
Weighted Average Energy Savings	64.0 MMBtu/yr
Weighted Average Incremental Cost	296 $2010
Product CCE	4.24 $2010

APPENDIX 3. DATA SOURCES FOR COST OF CONSERVED ENERGY CALCULATIONS

Table A.3.1. Data Sources for Residential Appliance Groups

Reference Title and Reference	Source Location By Category*			
	UEC	Price	Shipments	Lifetime
Refrigeration Equipment				
Refrigerator, Refrigerator-Freezer and Freezers Rulemaking Technical Support Document, USDOE (2010h)	Table 7.6.x**	Table 8.4.x		9.3.1.1 & 9.3.2.1
Refrigerator, Refrigerator-Freezer and Freezers NOPR National Impact Analysis Spreadsheet, USDOE (2010g)			'Historical Shipment & Market Share'	
Compact Refrigerator, Refrigerator-Freezer and Freezers NOPR National Impact Analysis Spreadsheet, USDOE (2010d)			'Historical Shipment'	
Room Air Conditioners				
Residential Clothes Dryers and Room Air Conditioners Preliminary Technical Support Document, USDOE (2010l)	7.3.x	8.2.x		8.1.1
Residential Clothes Dryers and Room Air Conditioners Preliminary NIA for Room Air Conditioners, USDOE (2010k)			'Shipments Forecast'	
Water Heaters				
Residential Water Heater National Impacts Analysis Spreadsheet, USDOE (2010m)	'LCC Inputs'	<- Same	'Base Case Shipments'	'Retirement Function'
Furnaces & Boilers				
Residential Furnaces and Boilers NIA Main, USDOE (2008c)	'LCC Inputs'	<- Same	'Base Case Shipments'	'Retirement function'

Table A.3.1. (Continued)

Cooking Products				
Residential Cooking Products Final Rule National Impact Analysis Spreadsheet, USDOE (2009c)	'Efficiency and Price'	<- Same	'Base Case Shipment Elec Cooking'	'Lifetime'
Residential Cooking Products Final Rule National Impact Analysis Spreadsheet: Including Microwave Ovens, USDOE (2009b)	'Input and Summary'	<- Same	'Base Case'	'Lifetime'
Residential Cooking Products Technical Support Document, USDOE (2009d)		Table 8.2.x		
Central Air Conditioning & Heat Pumps				
Residential Central Air Conditioners and Heat Pumps Preliminary Technical Support Document, USDOE (2010j)	Table 7.11.x	Table 8.2.x		Table 8.1.1
Residential Central AC/HP National Impact Analysis Spreadsheet, USDOE (2010i)			'Historical Shipments'	
Fluorescent Lamp Ballasts				
Fluorescent Lamp Ballasts Preliminary Technical Support Document, USDOE (2010f)	Table 6.3.x	Table 8.5.x		
Fluorescent Lamp Ballasts Preliminary National Impact Analysis, USDOE (2010e)			'Assumptions'	<- Same

* All single-quoted titles are the worksheet in the given analysis spreadsheet.

** The 'x' at the end of a table location indicates a range of tables all or most of which contain the desired values.

Table A.3.2. Data Sources for Commercial Building Appliance Groups

Reference Title and Reference	Source Location By Category*			
	UEC	Price	Shipments	Lifetimes
Distribution Transformers				
Distribution Transformer Final Rule National Impact Analysis Spreadsheet, USDOE (2007a)	LCC Data By Product Class' combining Load & No-Load Losses		'Market Share'	'Lifetime'
Distribution Transformer Final Rule Technical Support Document, USDOE (2007b)		Table 8.7.3		
Commercial Unitary Air Conditioners and Heat Pumps				
Commercial Unitary Air Conditioners and Heat Pumps ANOPR Technical Support Documents, USDOE (2003c)	Table 10.2.x**	Table 8.2.x		
Commercial Unitary Air Conditioners and Heat Pumps ANOPR Life Cycle Cost Analysis Spreadsheet (Tariff), USDOE (2003a)				'Lifetime'
Commercial Unitary Air Conditioners and Heat Pumps ANOPR National Energy Savings Spreadsheet, USDOE (2003b)			'65-135 Stock' & '135 -240 Stock'	
Packaged Terminal Air Conditioners and Heat Pumps				
Packaged Terminal Air Conditioners and Heat Pumps Final Rule Life-Cycle Cost Spreadsheet, USDOE (2008a)				'Lifetime'
Packaged Terminal Air Conditioners and Heat Pumps Final Rule National Impact Analysis Spreadsheet, USDOE (2008b)	'Equipment Parameters'	<- Same	'Shipments'	

Table A.3.2. (Continued)

Reference Title and Reference	Source Location By Category*			
	UEC	Price	Shipments	Lifetimes
High-Intensity Discharge Lamps				
High-Intensity Discharge Lamps Proposed Determination National Energy Savings Spreadsheet, USDOE (2004b)			'MV Lamp Shipments by Wattage' & '175W MV Applications' & '400W MV Applications'	
High-Intensity Discharge Lamps Proposed Determination Life-Cycle Cost Spreadsheet, USDOE (2004a)	Each product class has a worksheet which includes the UEC, price	<- Same		<- Same
Boilers				
ASHRAE Equipment (Boilers) Final Rule Technical Support Documents, USDOE (2010c)	Table 7.3.5	Table 7.3.4	Table 6.4.1	Section 5.6.2.4
Fluorescent Lamp Ballasts				
Fluorescent Lamp Ballasts Preliminary Technical Support Document, USDOE (2010f)	Table 6.3.x	Table 8.5.x		
Fluorescent Lamp Ballasts Preliminary National Impact Analysis, USDOE (2010e)			'Summary Shipments'	'Assumptions'

* All single-quoted titles are the worksheet in the given analysis spreadsheet.

** The 'x' at the end indicates a range of tables all or most of which contain the desired values.

*** The prices were calculated using the tables for base prices, incremental costs, markup, and installation cost for specifically 800 or 3000 kBtu/h boilers.

REFERENCES

ASAP (2011). Appliance Standards Awareness Project. *Federal Standards* table, Website: http://www.standardsasap.org/federal.htm.

BEA (2010). Bureau of Economic Analysis *Table 1.1.9. Implicit Price Deflators for Gross Domestic Product.* Available at www.bea.gov /national/nipaweb.

Caracino (2010). J. Caracino, M. Nevius, and L. Fesler, 2010. *State of the Efficiency Program Industry, 2009 Expenditures, Impacts & 2010 Budgets*. Consortium for Energy Efficiency December 10, 2010.

CEC (2005). California Energy Commission. *Implementing California's Loading Order for Electricity Resources*, Staff Report, CEC-400-2005-043, July 2005 http://www.energy.ca.gov/2005publications/CEC-400-2005-043/CEC-400-2005-043.PDF.

CEC (2009). California Energy Commission. *2009 Appliance Efficiency Rulemaking Staff Report – Phase I, Part C*. Docket # 09-AAER-IC. CEC-400-2009-024. September 2009.

CEE (2010). California Energy Commission. *State of the Efficiency Program Industry: 2009 Expenditures, Impacts & 2010 Budgets*. IN EFFICIENCY, C. F. E. (Ed.) Boston.

Display Search (2010). Display Search, *Quarterly Global TV Shipment and Forecast Report, 1st Quarter, 2010*. April 2010.

EIA (2003). Energy Information Administration. *2003 Commercial Building Energy Consumption Survey 2003. DOE.*

EIA (2005). Energy Information Administration. *2005 Residential Energy Consumption Survey. DOE.*

EIA (2010). Annual Energy Outlook 2010, Energy Information Administration.

EPA (2010) U.S. Environmental Protection Agency, "2009 Annual Report: Energy Star and Other Climate Protection Partnerships". December 2010. http://www.energystar.gov/ia/partners/publications/pubdocs/2009%20CPP D%20Annual%20Rep ort.pdf.

Federal Register (2010). Code of Federal Regulations: Title 10 - Energy. Chapter II – Department of Energy. Subchapter D – Energy Conservation. Part 430 – Energy Conservation Program for Consumer Products. Subpart A - General Provisions. IN GPO.GOV (Ed.).

Gold (2011). R. Gold, S. Nadel, S. Laitner and A. deLasky. "*Appliance and Equipment Efficiency Standards: A money maker and Job Creator.*" American Council for an Energy Efficient Economy / Appliance

Standards Awareness Project Report Number ASAP-8/ACEEE- A111. Available at http://www.aceee.org/research-report/a111.

IEA (2001). International Energy Agency. *Things that go Blip in the Night – Standby Power and how to Limit It*. OECD. 2001.

IEA (2006). International Energy Agency. *Light's Labour's Lost*, OECD, Paris.

Kushler (2004). Kushler, York and Witte. 2004. *Five Years In: An Examination of the First Half-Decade of Public Benefit Policies*. American Council for an Energy Efficient Economy.

McNeil (2008). M.A. McNeil, V.E. Letschert and S.A. de la Rue du Can. *Global Potential of Energy Efficiency Standards and Labeling Programs*. LBNL/CLASP. 2008. LBNL 760E.

USDOE (2003a). Commercial Unitary Air Conditioners and Heat Pumps ANOPR Life Cycle Cost Analysis Spreadsheet (Tariff-Based Electricity Prices). USDOE.

USDOE (2003b). Commercial Unitary Air Conditioners and Heat Pumps ANOPR National Energy Savings Spreadsheet. USDOE.

USDOE (2003c). Commercial Unitary Air Conditioners and Heat Pumps ANOPR Technical Support Documents. USDOE.

USDOE (2004a). High-Intensity Discharge Lamps Proposed Determination Life-Cycle Cost Spreadsheet. USDOE.

USDOE (2004b). High-Intensity Discharge Lamps Proposed Determination National Energy Savings Spreadsheet. USDOE.

USDOE (2007a). Distribution Transformer Final Rule National Impact Analysis Spreadsheet. USDOE.

USDOE (2007b). Distribution Transformer Final Rule Technical Support Document. USDOE. USDOE (2008a). Packaged Terminal Air Conditioners and Heat Pumps Final Rule Life-Cycle Cost Spreadsheet. USDOE.

USDOE (2008b). Packaged Terminal Air Conditioners and Heat Pumps Final Rule National Impact Analysis Spreadsheet. USDOE.

USDOE (2008c). Residential Furnaces and Boilers NIA Main. USDOE.

USDOE (2009a). Residential Cooking Products Technical Support Document. USDOE. USDOE (2009b). Residential Cooking Products Final Rule National Impact Analysis Spreadsheet: Including Microwave Ovens. USDOE.

USDOE (2009c). Residential Cooking Products Final Rule National Impact Analysis Spreadsheet: Not including Microwave Ovens. USDOE.

USDOE (2009d). Residential Cooking Products Technical Support Document. USDOE. USDOE (2009e) Impacts on the Nation of the Energy Independence and Security Act of 2007 - Technical Support Document. USDOE.

USDOE (2010a). Refrigerator, Refrigerator-Freezer and Freezers Rulemaking Technical Support Document. USDOE.

USDOE (2010b). Residential Clothes Dryers and Room Air Conditioners Preliminary Technical Support Document. USDOE.

USDOE (2010c). ASHRAE Equipment (Boilers) Final Rule Technical Support Documents. USDOE.

USDOE (2010d). Compact Refrigerator, Refrigerator-Freezer and Freezers NOPR National Impact Analysis Spreadsheet USDOE.

USDOE (2010e). Fluorescent Lamp Ballasts Preliminary National Impact Analysis. USDOE.

USDOE (2010f). Fluorescent Lamp Ballasts Preliminary Technical Support Document. USDOE.

USDOE (2010g). Refrigerator, Refrigerator-Freezer and Freezers NOPR National Impact Analysis Spreadsheet. USDOE.

USDOE (2010h). Refrigerator, Refrigerator-Freezer and Freezers Rulemaking Technical Support Document. USDOE.

USDOE (2010i). Residential Central AC/HP National Impact Analysis Spreadsheet USDOE. USDOE (2010j). Residential Central Air Conditioners and Heat Pumps Preliminary Technical Support Document. USDOE.

USDOE (2010k). Residential Clothes Dryers and Room Air Conditioners Preliminary NIA for Room Air Conditioners. USDOE.

USDOE (2010l). Residential Clothes Dryers and Room Air Conditioners Preliminary Technical Support Document. USDOE.

USDOE (2010m). Residential Water Heater National Impacts Analysis Spreadsheet USDOE.

End Notes

[1] World Energy Outlook 2010 Appendix A "Current Policies Scenario"

[2] Levelized capital cost of $74.6/MWh for "Advanced Coal" plants according to Energy Information Administration, Annual Energy Outlook 2011.

[3] DOE has set a standard implying heat pump technology for only the largest capacity water heaters, citing market barriers for the bulk of the market. As this technology becomes more prevalent, wider penetration may become achievable.

[4] Incandescent lamps show no savings in year 2030 itself, because they are forecast to be phased out by that year in the base case scenario.

[5] It should be emphasized that efficiency technology costs are generally decreasing, especially after a standard creates a mass market for high-efficiency equipment. An update of the data may therefore reveal additional cost- effective improvement.

[6] See http://www1.eere.energy.gov/buildings/appliance_standards/history.html.

[7] For simplicity the high efficiency scenario assumes 100% of the market will reach the target level in 2015, a structure that closely resembles minimum efficiency performance standards. In the later years of the forecast, the scenario is not highly sensitive to the details of the market transformation.

[8] BUENAS is described completely in McNeil, M.A., V.E. Letschert and S.A.De la Rue du Can (2008). *Global Potential of Energy Efficiency Standards and Labeling Programs*. LBNL 760E.

[9] More information about the LEAP platform may be found at http://www.energycommunity.org

[10] This depends somewhat on the lifetime of the product. For refrigerators we may assume a 15 - year lifetime, but some refrigerators may last 20 years, so the turnover of the stock may not be complete by 2030.

[11] Other metrics such as life cycle cost and payback period establish cost effectiveness, but are not easily compared across disparate technologies and end uses.

[12] In this paper, 'appliances' refers to a broad category of energy-consuming equipment, including lighting and HVAC equipment. An 'appliance' is defined here as a category of equipment used for a general purpose, such as lighting, refrigeration or heating. We distinguish this from the term 'product class', which we take to be a subset of an appliance group. For example, a top-mounted refrigerator-freezer is a product class within an appliance group called 'refrigerators'.

[13] Data for commercial building equipment and equipment used in industrial facilities have also been published as a result of the DOE program, but are not presented here.

[14] Maintenance and repair costs are included by DOE for products where these are significant. They are excluded here for simplicity and consistency.

[15] U.S. Department of Energy office of Energy Efficiency and Renewable Energy – *Building Energy Data Book 2010*, available at http://buildingsdatabook.eren.doe.gov.

[16] These data have since been updated, but were not publically available at the time of our analysis.

[17] Heat pump water heaters are already common in some other countries, notably Japan, China and Australia.

[18] 217 TWh of primary energy, using a site to source factor of 3.23 in 2010, compared to 14.4 quads total primary energy.

[19] The California Energy Commission recently found that television technology could be improved to meet its new minimum efficiency requirement at no additional cost (CEC 2009).

[20] 111 TWh of primary energy, using a site to source factor of 3.23 in 2010, compared to 14.4 quads total primary energy.

[21] Additional products were evaluated for cost-effectiveness with the result that no cost-effective improvements were possible beyond the current standard. These were removed from further consideration, and are not included in the Appendix.

[22] DOE appliance standards analyses typically use *marginal prices,* which represent the price of the last unit of energy paid, which can differ from average prices due to the structure of the tariff schedule. Strictly speaking, marginal prices are more appropriate than average prices when characterizing efficiency improvements. In addition, *regional prices* are also sometimes used for appliances with a large degree of regional variation in terms of product class share and energy consumption. For simplicity, neither of these refinements are considered here.

[23] It would be possible in principle to increase the efficiency savings potential by requiring only that the weighted average CCE be lower than the utility price. We chose the more stringent requirement that each product class be individually less than the utility price, since regulations generally apply at the product class level.

[24] DOE also found heat pump water heaters to be an appropriate MEPS target, but only for large capacity units, which account for a small fraction (around 10%) of the residential market.

[25] It is possible to model, for example, the reduction of sales or fuel switching resulting from price increases associated with efficiency regulations. This effect is not captured in BUENAS.

[26] The exception is the phase-out of incandescent lamps, which begins in 2012 in the *Business Case*.

[27] DOE has set a standard implying heat pump technology for only the largest capacity water heaters, citing market barriers for the bulk of the market. As this technology becomes more prevalent, wider penetration may become achievable.

[28] Incandescent lamps show no savings in year 2030 itself, because they are forecast to be phased out by that year in the base case scenario.

[29] It should be emphasized that efficiency technology costs are generally decreasing, especially after a standard creates a mass market for high-efficiency equipment. An update of the data may therefore reveal additional cost- effective improvement.

[30] World Energy Outlook 2010 Appendix A "Current Policies Scenario"

[31] Levelized capital cost of $74.6/MWh for "Advanced Coal" plants according to Energy Information Administration, Annual Energy Outlook 2011.

In: Business Case Scenarios for Energy Efficiency ISBN: 978-1-62948-076-3
Editor: Sean C. Byrne

Chapter 2

BUSINESS CASE FOR ENERGY EFFICIENCY IN SUPPORT OF CLIMATE CHANGE MITIGATION, ECONOMIC AND SOCIETAL BENEFITS IN CHINA*

Michael A. McNeil, Nicholas Bojda, Jing Ke, Yining Qin, Stephane de la Rue du Can, David Fridley, Virginie E. Letschert and James E. McMahon

EXECUTIVE SUMMARY

This study seeks to provide policymakers and other stakeholders with actionable information towards a road map for reducing energy consumption cost-effectively. We focus on individual end use equipment types (hereafter referred to as *appliance groups*) that might be the subject of policies - such as labels, energy performance standards, and incentives - to affect market transformation in the short term, and on high-efficiency technology options that are available today.

As the study title suggests, the high efficiency or *Business Case* scenario is constructed around a model of cost-effective efficiency improvement. Our analysis demonstrates that a significant reduction in energy consumption and

---

* This is an edited, reformatted and augmented version of Lawrence Berkeley National Laboratory, dated August 18, 2011.

emissions is achievable at *net negative cost*, that is, as a profitable investment for consumers. Net savings are calculated assuming no additional costs to energy consumption such as carbon taxes. Savings relative to the base case as calculated in this way is often referred to as “economic savings potential”.

Chinese energy demand has grown dramatically over the last few decades. While heavy industry still plays a dominant role in greenhouse gas emissions, demand from residential and commercial buildings has also seen rapid growth in percentage terms.

In the residential sector this growth is driven by internal migration from the countryside to cities. Meanwhile, income in both urban and rural subsectors allows ownership of major appliances. While residences are still relatively small by U.S. or European standards, nearly all households own a refrigerator, a television and an air conditioner. In the future, ownership rates are not expected to grow as much as in other developing countries, because they are already close to saturation. However, the gradual turnover of equipment in the world’s largest consumer market provides a huge opportunity for greenhouse gas mitigation. In addition to residences, commercial floor space has expanded rapidly in recent years, and construction continues at a rapid pace. Growth in this sector means that commercial lighting and HVAC will play an increasingly important role in energy demand in China.

The outlook for efficiency improvement in China is encouraging, since the Chinese national and local governments have implemented significant policies to contain energy intensity and announced their intention to continue and accelerate these.

In particular, the Chinese appliance standards program, first established in 1989, was significantly strengthened and modernized after the passage of the Energy Conservation Law of 1997. Since then, the program has expanded to encompass over 30 equipment types (including motor vehicles). The current study suggests that, in spite of these efforts, there is significant savings to be captured through wide adoption of technologies already available on the Chinese market. Potential impacts of this adoption include:

Energy savings:

- 490 billion kWh per year in 2020
- 1010 billion kWh per year in 2030
- A total of 10,000 billion kWh cumulatively through 2030

Cumulative greenhouse gas emissions mitigation:

- 7730 million metric tons of CO_2 through 2030
- 26 million metric tons of SO_2 through 2030
- 33 million metric tons of NO through 2030
- 1650 metric tons of mercury through 2030

Financial impacts to consumers through 2030:

- Equipment investment of 350 billion USD
- Energy bill savings of 1200 billion USD
- Net savings of 846 billion USD

The approach of the study is to assess the impact of short-term actions on long-term impacts. "Short-term" market transformation is assumed to occur by 2015, while "long-term" energy demand reduction impacts are assessed in 2030. In the intervening years, most but not all of the equipment studied will turn over completely. Early in 2011, the Chinese government announced a plan to reduce carbon dioxide emissions intensity (per unit GDP) by 16% by 2015 as part of the 12th five year plan. These targets are consistent with longer term goals to reduce emissions intensity 40-45% relative to 2005 levels by 2020. The efforts of the 12th FYP focus on short-term gains to meet the four-year targets, and concentrate mainly in industry. Implementation of cost-effective technologies for all new equipment in the buildings sector thus is largely complementary to the 12th FYP goals, and would provide a mechanism to sustain intensity reductions in the medium and long term. The 15-year time frame is significant for many products, in the sense that delay of implementation postpones economic benefits and mitigation of emissions of carbon dioxide. Such delays would result in putting in place energy-wasting technologies, postponing improvement until the end of their service life, or potentially resulting in expensive investment either in additional energy supplies or in early replacement to achieve future energy or emissions reduction targets.

The *Business Case* concentrates on technologies for which cost-effectiveness can be clearly demonstrated. The appliance groups studied are:

Residential End Uses
Incandescent Lamps
Refrigerators

Residential AC (Including Heat Pumps)
Washing Machines
Fluorescent Ballasts
Electric Water Heaters
Standby Power
Induction Ranges
Gas Ranges
Gas Water Heaters
Gas Boilers

Commercial and Industrial End Uses
Commercial Room AC
Commercial Lighting
Commercial Gas Boilers (Space Heating)
Commercial Gas Water Heaters
Industrial Motors
Distribution Transformers

Energy savings and greenhouse gas emissions mitigation for these appliance groups are summarized in Table ES-1.

Since the study includes only appliance groups for which cost-effectiveness can be clearly demonstrated, the benefits determined represent only a subset of the economy-wide potential. Specifically, transportation end uses and industrial processes technologies are not covered, because data sufficient to include them were not possible to collect within the scope of the research. Likewise, the study does not include system approaches such as smart grids. These approaches to efficiency may have important impacts but the calculation of costs and benefits is not as straightforward as for individual pieces of equipment. In addition, the technologies analyzed represent a snapshot of what is currently on the market. Technological innovations are certain to occur over the coming decades, and these will likely present new opportunities for efficiency improvement, and exert downward pressure on costs.

Efficiency measures are determined to be cost-effective if the cost of conserved energy associated with them is less than the consumer's energy price, that is, the amount saved in energy bills is greater than the initial investment. The *Business Case* scenario is generated by identifying the *maximum efficiency improvement for which cost of conserved energy is lower than utility energy prices* (projected to 2015). The relative contribution to cumulative emissions for each appliance group is shown in Figure ES-1.

Table ES-1. Energy Savings and Pollutant Mitigation by Appliance Group

Appliance Group	Final Energy Savings			Emissions Mitigation					
	In 2020	In 2030	Through 2030	In 2020	In 2030	Through 2030			
	TWh			mt CO_2		mt CO_2	mt SO_2	mt NO	t Hg
Gas Water Heaters	118	255	2448	24	52	494	6.2	8.1	403.9
Distribution Transformers	57	148	1273	60	146	1294	3.2	4.2	210.1
Commercial Lighting	63	121	1249	66	119	1277	3.2	4.1	206.1
Residential AC (Inc. HP)	57	90	1020	60	89	1041	2.6	3.4	168.3
Refrigerators	49	100	991	51	98	1009	2.5	3.3	163.5
Gas Boilers (Res.)	26	75	607	5	15	123	1.5	2.0	100.1
Commercial Gas Boilers (Space Heating)	32	49	567	7	10	114	1.4	1.9	93.5
Washing Machines	25	50	505	27	49	514	1.3	1.7	83.3
Industrial Motors	14	40	338	15	40	343	0.9	1.1	55.8
Incandescent Lamps	10	38	282	11	38	286	0.7	0.9	46.5
Standby Power	12	20	210	12	19	214	0.5	0.7	34.6
Electric Water Heaters	12	17	203	12	17	207	0.5	0.7	33.5
Commercial Room AC	11	14	185	11	14	190	0.5	0.6	30.5
Fluorescent Ballasts (Res.)	7	9	108	7	9	110	0.3	0.4	17.7
Induction Ranges	1	2	16	1	2	16	0.0	0.1	2.6
Total	494	1028	10001	368	717	7234	26	33	1650

Several conclusions can be drawn from Table ES-1 and Figure ES-1. First, emission reduction potential is well distributed among end uses and sectors. The largest potential exists for commercial lighting, residential air conditioning and distribution transformers, each of which could provide over 100 TWh of electricity demand reduction in 2030 and over 1000 mt CO_2 over the forecast period. Savings in commercial lighting is possible due to the prevalence of magnetic ballasts for fluorescent lamps, which can be replaced with electronic ballasts. Air conditioning savings includes reduction of heating energy from reversible models (heat pumps). Finally, distribution transformers show significant savings because of the scale of distribution losses at the national level, even though total efficiency improvement is small on a percentage basis. Much of the remaining mitigation potential comes from major appliances such as refrigerators and washing machines. Significant savings can be gained from replacement of gas and electric water heaters with solar water heating systems, which already have a significant market share in China.

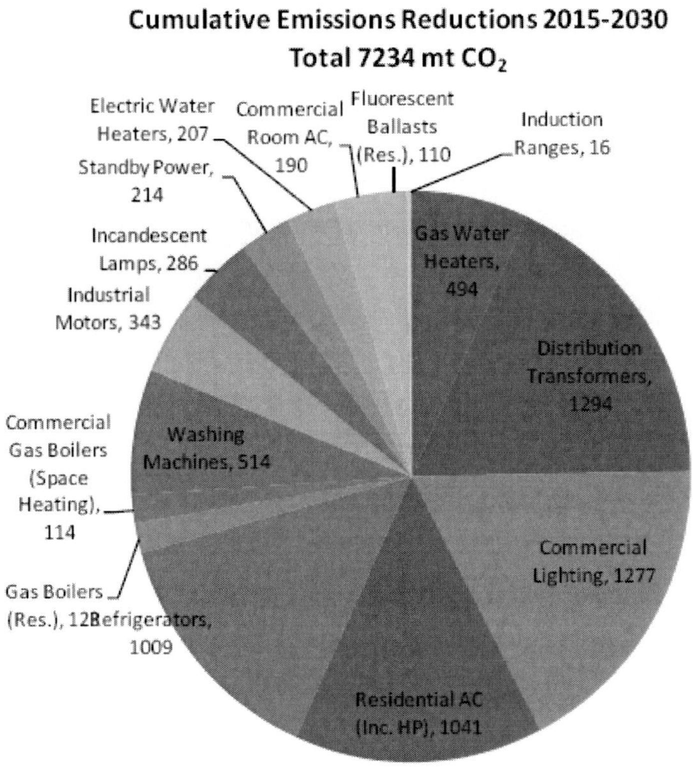

Figure ES-1. Contribution to Cumulative CO_2 Emissions Reductions 2010-2030.

The analysis shows that cost-effective efficiency improvement could yield very significant financial benefits to Chinese consumers. Table ES-2 shows positive net savings for all appliance groups, which is not surprising, since the target efficiency levels were constructed to be cost-effective. The table shows that cost-effective efficiency improvements require an investment of 354 billion USD over the next 20 years, but these investments will return over more than three times as much over the same period, for a net savings of 850 billion dollars, or of order of six hundred dollars per capita. The present value of net savings is 554 billion USD assuming a discount rate of 3%, and 328 billion USD with a 7% discount rate.

Of the appliance groups studied, residential room air conditioners require the largest investment at 97 billion USD, but provide a payoff of 148 billion USD. Refrigerators generate similar savings, but with only half of the investment. Phasing out incandescent lamps is extremely cost effective compared to other appliance groups, with a payoff of nearly 10 times as high as the required investment. Finally, distribution transformers are by far the best investment, with a savings of almost 10 times the investment of 25 billion USD.

Table ES-2. Cumulative Financial Impacts of Efficiency Improvement

Appliance Group	Cumulative Financial Impacts				
	Cost	Savings	Net Savings	NPV @ 3% DR	NPV @ 7% DR
	Billions USD				
Commercial Lighting	25.0	242	217	145	88.0
Distribution Transformers	7.7	220	212	137	79.7
Refrigerators	45.6	144	98.5	64.3	37.8
Gas Water Heaters	52.1	116	63.8	41.5	24.4
Residential AC (Inc. HP)	96.6	148	51.8	34.2	20.4
Industrial Motors	11.0	58.4	47.4	30.5	17.6
Incandescent Lamps	4.6	41.0	36.4	23.2	13.2
Standby Power	1.9	30.5	28.6	18.9	11.3
Washing Machines	47.9	73.4	25.5	16.7	9.8
Commercial Gas Boilers (Space Heating)	11.8	31.7	19.9	13.4	8.3
Commercial Room AC	21.4	35.9	14.5	9.9	6.2
Electric Water Heaters	12.4	29.6	17.1	11.3	6.8
Gas Boilers (Res.)	16.1	28.7	12.6	8.1	4.7
Fluorescent Ballasts (Res.)	6.4	15.6	9.2	6.1	3.7
Induction Ranges	0.7	2.3	1.6	1.0	0.6
Total	354	1200	846	554	328

1. Introduction

Some recent examples of studies that have identified potential energy savings from energy efficiency improvements include:

- A Lawrence Berkeley National Laboratory (LBNL) report, *Impacts of China's Current Appliance Standards and Labeling Program to 2020*, estimated cumulative savings of 1143 TWh (9% of cumulative residential electricity use) from 2000 to 2020 for China's standards and labeling programs for eleven key energy-consuming products (Fridley 2007)

- Another 2010 LBNL study estimated cumulative 2009-2030 electricity savings of 9503 TWh under a scenario of regularly scheduled appliance standards revisions and cumulative savings of 5450 TWh for a subset of products that achieves international best practice (Zhou 2010a).
- The China National Institute of Standardization (CNIS)'s 2010 White Paper estimated 198.7 TWh of annual electricity savings in 2030, or about 6.3% of total annual electricity consumption from China's appliances and equipment standards and labeling programs covering a subset of key products subject to standards (CNIS 2010)

This study seeks to provide policymakers and other stakeholders with actionable information towards a road map for reducing energy consumption in the most cost-effective way. A major difference between the current study and some others is that we focus on individual equipment types that might be the subject of policies - such as labels, energy performance standards, and incentives - to affect market transformation in the short term, and on high-efficiency technology options that are available today.

The approach of the study is to assess the impact of short-term actions on long-term impacts. "Short term" market transformation is assumed to occur by 2015, while "long-term" energy demand reduction impacts are assessed in 2030. In the intervening years, most but not all of the equipment studied will turn over completely. Early in 2011, the Chinese government announced a plan to reduce carbon dioxide emissions intensity (per unit GDP) by 16% by 2015 as part of the 12th five year plan. These targets are consistent with longer term goals to reduce emissions intensity 40-45% relative to 2005 levels by 2020. The efforts of the 12th FYP focus on short-term gains to meet the four-year targets, and concentrate mainly in industry. Implementation of cost-effective technologies for all new equipment in the buildings sector thus is largely complementary to the 12th FYP goals, and would provide a mechanism to sustain intensity reductions in the medium and long term. The 15-year time frame is significant for many products however, indicating that delay of implementation postpones impacts such as net economic savings and mitigation of emissions of carbon dioxide. Such delays would result in putting in place energy-wasting technologies, postponing improvement until the end of their service life, or potentially resulting in expensive investment either in additional energy supplies or in early replacement to achieve future energy or emissions reduction targets.

1.1. Policies and Programs to Encourage Efficiency

Chinese energy demand has grown dramatically over the last few decades. While heavy industry still plays a dominant role in greenhouse gas emissions, demand from residential and commercial buildings has also seen rapid growth in percentage terms. In the residential sector this growth is driven by internal migration from the countryside to cities. Meanwhile, household income in both subsectors allows ownership of major appliances. While residences are still relatively small by U.S. or European standards, nearly all households own a refrigerator, a television and an air conditioner. In the future, ownership rates are not expected to grow as much as in other developing countries, because they are already close to saturation. However, the gradual turnover of equipment in the world's largest consumer market provides a huge opportunity for greenhouse gas mitigation. In addition to residences, commercial floor space has expanded rapidly in recent years, and construction continues at a rapid pace. Growth in this sector means that commercial lighting and HVAC will play an increasingly important role in energy demand in China.

The outlook for efficiency improvement in China is encouraging, since the Chinese national and local governments have implemented significant policies to contain energy intensity and announced their intention to continue and accelerate these. In particular, the Chinese appliance standards program, first established in 1989, was significantly strengthened and modernized after the passage of the Energy Conservation Law of 1997. Since then, the program has expanded to encompass over 30 equipment types (including motor vehicles). The current study suggests that, in spite of these efforts, there is significant savings to be captured through wide adoption of technologies already available on the Chinese market.

Efficiency Standards

China first enacted minimum energy performance standards in 1989 for eight residential products including refrigerators, clothes washers, air-conditioners, fans, rice cookers, televisions, radio receivers and recorders, and electric irons. However, the energy efficiency requirement at the time was rather low and the monitoring and enforcement was almost inexistent. Since,

China has beneficiated from the assistance of international experts, notably the Collaborative Labeling and Appliance Standards Program (CLASP), to develop the necessary infrastructure capacity to develop and implement product standards (Zhou, 2008). In 1997, China promulgated the Energy Conservation Law which highlighted the importance of end-use energy efficiency and standards and label programs. As a result, energy efficiency standards were revised and extended to more products including fluorescent lamp ballasts, compact and linear fluorescent lamps. More recently, the Eleventh Five Year Plan (2006 to 2010) sets an ambitious energy efficiency objective of reducing energy consumption per unit of GDP by 20% by 2010. Standard and labeling programs are major piece of the plan to achieve the energy efficiency target. The close collaboration with international experts has propelled a very comprehensive and robust China's appliance standards and labeling program.

The National Development and Reform Commission (NDRC) is the main government institution responsible to develop and manage energy efficiency policy in China. AQSIQ (State Administration of Quality, Supervision, Inspection, and Quarantine) is responsible for developing the MEPS with administrative support from the Standardization Administration of China (SAC), and technical support from the China National Institute of Standardization (CNIS). (Fridley, 2007)

Further, China launched a voluntary label program in 1999, designed to identify and promote energy-efficient products. The program was managed by the China Standards Certification Center (CSC) under CNIS, but management of the label was transferred to the China Quality Certification Center (CQC) in 2008. It now covers over 50 products from over 300 participating manufacturers including: home appliances; consumer electronics; office equipment; lighting; and selected industrial equipment (Zhou, 2008).

In 2005, China launched a mandatory comparative energy information label for air conditioners, refrigerators, and clothes washers. The label includes five categories of efficiency, from 1 (more efficient) to 5 (less efficient). The program is managed by CNIS and was further extended in June 2008 to include gas water heaters, water chilling packages, three-phase induction motors, high-pressure sodium lamps and self-ballasted lamps (IEA, 2009). It now covers 26 products, and the label in some cases, such as consumer electronics, gas water heaters, and air conditioners, has been simplified to three categories from five.

Financial Incentives

China's NDRC and the Ministry of Finance currently provide subsidies to companies whose products reach a defined energy efficiency level. In April 2008, China launched the Financial Subsidies Fund for Promoting High Efficient Lighting Products Green Lights program to increase the use of energy-efficient light bulbs. The program is the first financial subsidy type of program used in China to promote the penetration of energy efficient products. Subsidies were offered to suppliers to provide a 30 percent discount on wholesale purchases and a 50 percent discount on retail sales. A total of 210 million subsidized CFLs were sold to consumers between 2008 and 2009, which had resulted in an estimated savings of 8.8 billion kWh of electricity each year, a reduction of 8.8 million tonnes of carbon dioxide per year. China planned to subsidize the price to the public of 150 million compact fluorescent light bulbs in 2010. (China Daily, 2010).

In May 2010, the Chinese government extended the incentive program to other products by allocating over 400 billion RMB (60 billion USD) annually to promote domestic demands for energy efficient products that include ACs, refrigerators, washers, TV, motors, and computer displays. For example, the central government offers subsidies of 500-850 RMB per unit for grade 1 AC products and 300-650 RMB per unit for grade 2 AC products. Local governments provide additional subsidy of 150 RMB for grade 1 AC units and 100 RMB for grade 2 units. (Wang, 2010).

ESCOs

In April 2010, the State Council, China's central government, issued a measure called "Opinions on Accelerating the Promotion of Energy Performance Contracting to Boost the Energy Service Industry," which provides new financial and tax incentives for energy service companies carrying out energy-performance contracting.

Utility Programs

Historically, China's DSM experience has largely focused on load management rather than energy efficiency, in part to address a massive power shortage. No mechanisms were set up to encourage utilities to design and implement energy efficiency programs. Recently, a few pilot projects developed with international collaboration in Jiangsu, Hubei, Shanghai, and Guangzhou have been generating significant electricity savings that has caught the attention of the Chinese top leadership (EF, 2006; Moskovitz, 2005). On November 2010, the China's central government enacted national energy efficiency regulations that require China's power grid companies to achieve energy savings equivalent to at least 0.3% in sales volume and 0.3% in maximum load compared with their previous year (Finamore, 2010). The new regulation came into effect on January 1, 2011. Possible sources of funding include: a rate surcharge applicable to all, rate surcharge differentiated by customer categories, and government budget allocations (Finamore, 2010). These new regulations have largely been inspired by the Energy Efficiency Resource Standards passed in many US States and notably in California. Responsibility for DSM implementation is placed on the shoulders of power grid companies by requiring these companies to invest in energy efficiency programs for their customers. Ongoing research and international collaboration are analyzing the implementation process for adopting effective DSM policies (Hu et al., 2005).

The experience in Chinese provinces already gives insight of the future development outcomes of the new regulations. In China, the term "DSM" generally relates to load management. Instead of DSM, the term Efficiency Power Plant (EPP) has been more wildly used in provinces to describe a virtual power plant that delivers "negawatts" through a portfolio of demand side management or energy efficiency projects.

An example of such an approach is the large-scale industrial EPP pilot in China's Jiangsu Province, developed under a California-Jiangsu partnership on energy efficiency improvements. In 2004, Natural Resources Defense Council (NRDC), along with the Energy Foundation (EF) and the Regulatory Assistance Project (RAP) worked with China's national government and Jiangsu Province to help launch China's first large-scale DSM pilot program in Jiangsu. Furthermore an agreement on DSM cooperation was signed between the California Public Utilities Commission, the California Energy Commission, and the Jiangsu Economic and Trade Commission in 2005. The joint work has helped Jiangsu build a large-scale industrial EPP program equivalent to a 300MW power plant during the period from 2005 to 2007. The implemented energy efficiency measures have helped Jiangsu reduce 580 megawatts of peak load, save 2 TWh of electricity annually, and reduce CO2 emissions by 1.88 million tons (Shen et, 2009). These initial efforts laid the foundation for demonstrating in China the value of establishing large-scale DSM programs to meet environmental and energy goals.

In Hebei, the local government collects, on an annual basis, 0.01RMB/kWh from a surcharge called urban construction fee that is levied on every customer's electric utility bill and utilizes the money for supporting EPP programs.

Shandong has implemented an energy quota system covering 20 industries and 52 products manufactured in the province. The provincial government has set energy use (electricity and other fuels) quota levels. Consumers who exceed the quota level pay a substantial surcharge, as much as 400% of the energy price. The surcharge is paid to the Shandong Energy Conservation Supervision Center and is deposited in a special fund to be used for energy efficiency. (Moskovitz et al., 2007).

There is a growing trend in China to use financial incentives to foster the market penetration of energy efficient products. The national government already implemented subsidies to manufacturers of energy efficient appliances and a few provincial governments have developed their own target for utility to promote energy efficiency at the end use level. The new regulations that set an obligation for all utility companies promise to bring significant energy savings over the next years.

2. Energy Demand Scenarios

As the study title suggests, the high efficiency or *Business Case* scenario is constructed around a model of cost-effective efficiency improvement. The point of the study is to demonstrate that a significant reduction in energy consumption and emissions is achievable at a net negative cost, that is, as a profitable investment for society. There are a variety of ways of assessing costs and benefits to society. We chose to focus on the end user's perspective: costs in terms of additional retail equipment prices (capital investments); savings from reduced energy bills (operating costs). Only direct energy savings are included, without valuing non-energy benefits that may also accrue (comfort, productivity, health). Finally, the cost-benefit analysis is made without the elevated effective energy prices that could be implied by carbon taxes, carbon trading schemes or other policies. Savings relative to the base case as calculated in this way is often referred to as "economic savings potential".

A national-level high-efficiency scenario is constructed by assuming that market transformation to high-efficiency technologies will occur by 2015, which is judged to constitute the "short term" by the study, because it considers that five years is sufficient time to achieve market transformation through aggressive policies and stakeholder actions. The study does not model specific actions, which could include mandatory standards, voluntary labeling programs, voluntary agreements by manufacturers, utility demand-side-management programs and others[1].

The target efficiency level chosen is that which *maximizes efficiency while providing a net benefit to consumers*. This is to be contrasted with scenarios which maximize consumer payoff but not necessarily efficiency improvement, or those that include the best available technology ("max tech") without consideration of cost-effectiveness. Consumer cost-benefit analysis is evaluated in terms of cost of conserved energy. Cost of conserved energy (CCE) is the amortized incremental cost of equipment divided by annual energy savings. In other words, it's the additional annual capital investment needed to purchase high-efficiency equipment instead of baseline equipment, divided by the energy savings provided by the investment. This quantity, which has units of USD per unit energy, can be compared to prevailing energy prices to assess consumer cost-effectiveness. Technologies with a CCE less than forecast energy prices in 2015 are deemed cost-effective.

A few comments about whether this definition is optimistic or pessimistic are warranted. On one hand, high efficiency technologies are compared to the current baseline technology, even though there may already be a market for higher efficiency equipment, and the average efficiency of the market is constantly improving. This tends to underestimate the baseline forecast and overestimate savings. On the other hand, it likely underestimates the efficiency that will be achievable in a cost-effective way, first of all because technology costs are generally decreasing (according to technological learning rates) and the emergence of new technologies that may not be available for analysis. Therefore, there are two compensating effects not taken into account in the analysis. The results should therefore be taken as representative of the scale of potential improvement, not as a reliable prediction. The methodology is chosen to maximize concreteness and defensibility by relying on technologies that can be justified by actual cost data.

2.1. Literature Review

Some recent examples of studies that have identified potential energy savings from energy efficiency improvements include:

China

- China's appliance standards are estimated to have saved 1.08 EJ during 2006-2008, with refrigerators, air conditioners and televisions contributing the bulk of the savings. (Price et al. 2011)
- (Fridley 2008) estimates potential savings of 1.2 TWh in 2012 and 16 TWh by 2020 for energy labels on refrigerators in China.
- (Cheung 2008) describe the growth of China's energy efficiency industry, projecting spending of USD 300 billion over five years.
- (Aden 2010) uses lifecycle assessment to show that for buildings in the Beijing area, 80% of energy use and related emissions is due to operations, and about 20% due to materials.
- (Zhou 2010) provides an overview of China's policies on energy efficiency.

India

- (Delio 2009) estimates potential savings from energy efficiency across all sectors in India to be 183 TWh in five years.
- (De la Rue du Can 2009) provides both retrospective and prospective views of energy use in the residential and transport sectors of India.

United States

- The National Research Council report, *America's Energy Future,* in 2009 estimated potential cost-effective energy savings in the U.S. of about 20% in 2020 and about 30% in 2030, with the greatest potential in the buildings sector (National Research Council, *Limiting the Magnitude of Future Climate Change*, 2010).
- The American Physical Society report, *Energy Future: Think Efficiency* (2008) estimated 572 TWh of electricity savings in the residential sector in 2030, and about 30% savings for the building sector as a whole, all below the retail price of electricity energy.
- The U.S. Department of Energy's Appliance Standards Programs has conducted extensive studies for regulated product types (http://www1.eere.energy.gov/buildings/appliance_standards/), identifying economically justified and technologically feasible energy efficiency improvements.
- The Energy Information Administration annually publishes additional efficiency scenarios, e.g., high technology cases, in conjunction with the *Annual Energy Outlook* (http://www.eia.doe.gov/oiaf/aeo/).

2.2. Construction of the Energy Demand Scenarios

Any study that aims to project energy efficiency improvements from specific technologies must make the link between unit-level improvements and national impacts. The current study achieves this using LBNL's Bottom-Up Energy Analysis System (BUENAS). As the name suggests, BUENAS is a bottom-up technology-oriented model, rather than a top-down macroeconomic model[2]. BUENAS combines unit-level efficiency scenarios with a forecast of stock size and turnover to calculate national energy savings impacts through 2030. Unit level energy demand by baseline and "target" technologies are collected in a database that the model takes as inputs, and which define the base case and high efficiency scenarios. Growth of the stock (number of units operating) by 2030 is a function of economic and population growth.

The application of BUENAS to Chinese building end uses was implemented in collaboration with LBNL's China Energy Group and follows a somewhat different methodology than other regions in the 'global' version of the model. First of all, while urbanization is a driver variable in the global model, in the China model the sector division is more distinct, with separate assumptions and econometric parameters for urban and rural sectors. Second, since electricity access is basically universal in China, electrification is not an activity driver. Therefore, household income is the main driver of appliance ownership. The relationship between income growth and appliance ownership is determined for each sector using appliance ownership rates provided by the China's National Bureau of Statistics in its annual *China Statistical Yearbook.* The resulting version of BUENAS not only supports the current study, but constitutes a component in LBNLs "China Energy 2050" model. Further description of the modeling of the Chinese residential sector can be found in (Letschert 2009).

BUENAS uses minimum efficiency performance standards (MEPS) as a default policy, that is, it models a discrete change in the efficiency of equipment after a specific year. For the current study, we chose an implementation year of 2015, assuming that several years lead time are necessary between identification of efficiency targets, and making them mandatory.

Originally constructed as a global model, BUENAS covers a wide range of energy-consuming products, including most appliance groups generally covered by Energy Efficiency Standards and Labeling (EES&L) programs around the world. The global model covered the following appliance groups:

- *Residential Sector:* Lighting, Refrigerators, Air Conditioners, Fans, Washing Machines, Standby Power, Televisions, Electric Ovens, Space Heating and Water Heating.
- *Commercial Building Sector:* Lighting, Air Conditioning, Refrigeration, Ventilation, Office Products, Space Heating and Water Heating.
- *Industrial Sector:* Electric Motors.

For the purposes of the *China Business Case for Energy Efficiency*, many of the end uses needed for the analysis were present in BUENAS. However, many modifications were made. First, the *Business Case* model is dependent on an evaluation of cost-effectiveness. Therefore, appliance groups for which data were insufficient to permit this calculation were not included. On the other hand, some equipment types for which data were available were not included in the original model. In that case, these end uses were added. In the residential sector, all of the original end uses were covered for China except for fans and electric ovens. Due to the prevalence of air conditioners in China, residential fans do not account for a significant savings opportunity. Electric ovens were replaced by electric induction ranges and gas ranges, which are much more common in Chinese households. Residential space heating included both electric heat pumps and gas boilers. Electric room heaters were not covered. Also, much of residential heating in Chinese homes is district heat – while this end use may afford cost-effective savings, these savings are difficult to analyze in terms of direct consumer equipment costs and energy cost savings. Finally, televisions are not covered here. While efficiency improvements are certainly possible in television displays, this is a dynamic and rapidly evolving technology. Recent market trends include a massive shift to flat panel technology, with dramatic increases in screen size, along with market-driven efficiency improvements. Because of the dynamism in television technology, efficiency baselines and technology trends cannot be adequately forecast in order to provide definitive cost-effective improvement potential.

In the commercial sector, both central and room air conditioners were considered for cost-effective improvement, but a cost-effective target for central air conditioners was not found in the data, partially due to the less transparent nature of pricing for this type of equipment relative to residential equipment. This end use deserves further study than was in the scope of the current project. Fluorescent ballast efficiency improvement was considered for lighting, as were gas boilers for space heating, and gas water heaters.

Distribution transformers were added for this sector, but ventilation, refrigeration and office products were not considered due to lack of data.

The BUENAS model uses the Long Range Energy Alternatives Planning (LEAP) platform[3] to forecast energy consumption by end use from 2005 (base year) to 2030. The strategy of the model is to first forecast end use *activity*, which is driven by increased ownership of household appliances and growth in the industrial sector. The total stock of appliances can be modeled either according to an econometric diffusion model or according to unit sales forecasts, if available. Electricity consumption or *intensity* of the appliance stock is then calculated according to estimates of the baseline intensity of the prevailing technology in the local market. Finally, the total final energy consumption of the stock is calculated by modeling the flow of products into the stock and the marginal intensity of purchased units, either as additions or as replacements of old units. The high efficiency or "policy" scenario is created by the assumption of increased unit efficiency relative to the baseline starting in a certain year. For example, if the average baseline unit energy consumption (UEC) of new refrigerators is 450 kWh/year, but a MEPS taking effect in 2012 requires a maximum UEC of 350 kWh/year, the stock energy in the policy scenario will gradually become lower than that of the base case scenario due to increasing penetration of high-efficiency units under the standard. By 2030, the entire stock will generally be impacted by the standard[4]. Figure 1 shows the analytical structure of BUENAS.

The main outputs of BUENAS are base case energy consumption forecasts to 2030 by end use and energy, energy saving impacts of the modeled policy, and carbon dioxide emissions mitigation impacts. For this study, financial impacts were added to the model in a spreadsheet calculation.

For the residential sector, *activity* as modeled in Module 1 of the model is given by the stock of equipment, that is, the number of appliances installed and operating in Chinese households in a given year.

Once the number of residential products in each appliance group in each year is established, this number is multiplied by the *annual unit energy consumption* (UEC) to yield energy demand for the appliance group. *UEC* is the subject of Module 2 of BUENAS, and determines the efficiency scenario modeled. Determination of the baseline and efficiency scenario UEC is discussed in Section 3 below.

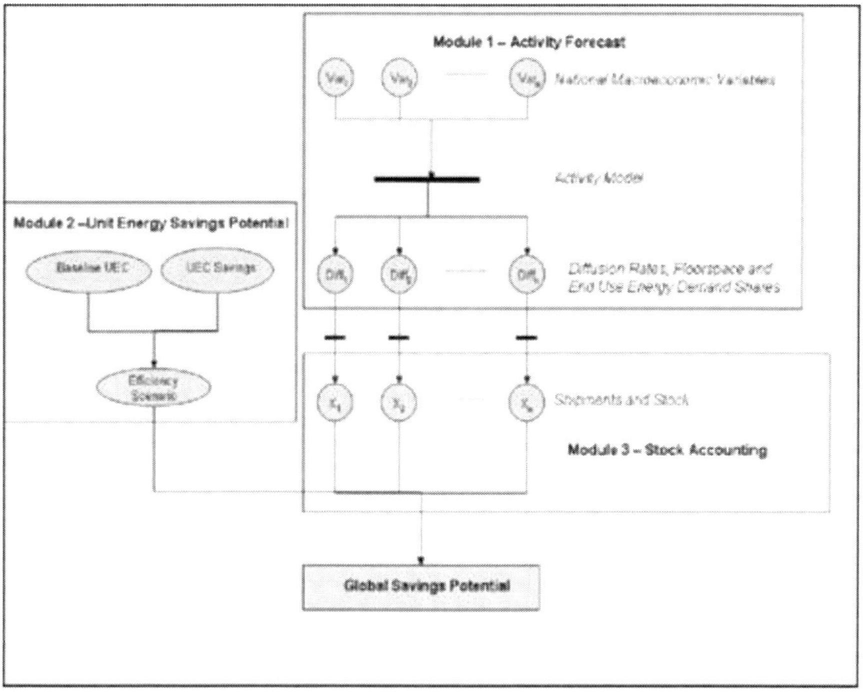

Figure 1. Structure of BUENAS.

Finally, Module 3 tracks the introduction of each year's cohort of appliances into the stock, taking account of growth in the market, equipment retirements, and replacements. Retirement and survival functions are derived from average lifetimes and assumed to have a distribution around the mean value. This shape of the retirement function is assumed to be that of a normal distribution centered around the mean lifetime by default, but takes the form of a more complicated function (Weibull distribution) if such a distribution is available. The survival function is given by:

$$Survival(age) = 1 - \int Retirement(age)$$

Using the retirement distribution, the model calculates the weighted average efficiency of the stock in each year. In the case of the high efficiency scenario, only a small fraction of the stock operates at high efficiency in the years immediately following the policy start date, but this fraction grows over time. The percentage of stock operating in 2030 that was installed after the policy start date is dependent on the assumed average lifetime of the product class.

Compared to the residential sector, energy demand in the commercial building sector is driven by a much wider variety of equipment types and follows distinct usage patterns depending on the type of building. For this reason, BUENAS models commercial buildings in an aggregate fashion, rather than at the level of individual appliances. The activity variable in this case is commercial building floor space. Commercial floor space projections for China are taken from LBNLs China Energy 2050 model (Zhou 2011). This model forecasts the growth in China's commercial building sector according to expected increases in tertiary sector employment and per-employee floor space. In Module 2, the commercial sector model uses aggregate energy intensity numbers for major appliance categories, such as lighting, space heating and air conditioning and refrigeration. In order to model energy demand and savings from efficiency improvement, we estimate the fraction of energy covered by individual technologies for which data are available. Energy and demand are thereby calculated from base year values of energy intensity according to a scaling factor.

3. Efficiency Improvement Potential – Cost-Benefit Analysis

Cost-effectiveness is defined in terms of cost of conserved energy, that is, how much the end user must pay in terms of annualized incremental equipment investment for each unit of energy saved by higher efficiency equipment. The formula for cost of conserved energy is

$$CCE = \frac{I \times q}{S} \qquad (1)$$

In this equation, I is the total additional investment needed to purchase high efficiency equipment rather than the baseline technology, and S is the resulting annual energy savings. The capital recovery factor q is given in turn by:

$$q = \frac{d}{(1-(1+d)^{-L})} \tag{2}$$

In this equation, d is the end user discount rate and L is the average lifetime of the equipment, in years. Defined in this way, I times q is an annual payment for an amortized capital investment. Cost of conserved energy is a convenient metric for comparison of cost-effectiveness of measures[5].

3.1. Equipment Data

The evaluation of cost-effectiveness of Chinese efficiency technologies for this study relied heavily on current retail price data taken from retail websites in China. Identification of efficiency for most products studied was facilitated by the mandatory Energy Label program, covering 26 product types, since most models were identified with a specific 1 to 5 efficiency level, where the "5" level typically is the minimum allowed efficiency and "1" is the most efficient level. Efficiency levels were combined with usage estimates from the Chinese National Institute of Standards (CNIS) and other sources to establish estimates of annual unit energy consumption. The result of this data collection is a database of prices and consumption level allowing evaluation of costs and benefits of substituting one model for another. Assumptions of baseline energy consumption, high efficiency levels and price data sources are shown in Table 1.

Incandescent Lamps

Replacement of incandescent lamps with compact fluorescent lamps (CFLs) or other technology such as LEDs is generally at the top of the list of attractive efficiency measures because of the large fractional savings (up to 60%) and the high degree of cost-effectiveness. Bans on incandescent lamps are also among the most popular efficiency policies globally. China has announced a phase-out of incandescent bulbs by 2018 according to NDRC. 6

Although this regulation is currently "on the books", we chose to include its effects as part of the *Business Case* scenario, rather than in the Base Case. In this case, the high-efficiency scenario is characterized by successful implementation and enforcement of the policy, rather than the establishment of it. We assume that the typical incandescent lamp in China is 60W and is operated for 2.3 hours per day on average, for an annual energy consumption of 53.4 kWh. An equivalent CFLs is assumed to use only 15W, or 12.6 kWh per year. We assume that a CFL lasts for 5 years, compared to only 1 year for incandescent.

Refrigerators

According to data from CNIS, refrigerators capacity has been growing in capacity over time, significantly increasing their energy consumption in Chinese households. At the same time, Chinese minimum efficiency performance standards and labels have contributed to a reduction of consumption than would otherwise have been the case. We assume that the baseline refrigerator corresponds to the China Energy Label Level 5, and that the market average refrigerator consumes 1.5 kWh per day, or 547.5 kWh per year. We assume that a refrigerator lasts 15 years.

Room Air Conditioners and Heat Pumps

Room air conditioners are the dominant technology for residential space cooling in China, with the most common configuration being split units. Room air conditioners, including reversible mode units (heat pumps) are used in both residential and commercial applications. In China, heat pumps enjoy a large market share, which we assume to be about 70% of the market. A revised minimum efficiency standard that went into effect in July 2010 reduced the label categories to three, with Level 3 being the mandatory minimum;; therefore we take this level as the baseline, and consider Level 1 as the high-efficiency level.

In order to calculate cooling and heating energy, we assume that the typical RAC unit is 1.125 kW and that it operates 4 hours per day every day during a 3 month cooling period. In commercial applications, we assume that the air conditioner is used for 8 hours per day in the same cooling season. The resulting annual baseline energy consumption estimates are 385 kWh per year for residential cooling and 770 kWh per year for commercial cooling.

Where the same unit is used for heating (residential sector only), we make the following regional assumptions:

- Northern Region: 20% of heat pump market - 6 hours per day for two weeks (shoulder period of heating season when district heating is turned off).
- Transition Region: 80% of the heat pump market - 6 hours per day for 3 months.

These assumptions result in a weighted average baseline consumption of 891 kWh per year for heating and cooling. The average lifetime of Room Air Conditioners is assumed to be 12 years.

Washing Machines

Two types of washing machines are popular in China - vertical axis washing machines and horizontal axis (front load) washing machines. The proportion of front-loading machines is increasing very quickly. From 2006 to 2010, the sales share of this product class increased from 12.7% to 58.4%[7]. Therefore we assume that this will be the dominant technology in 2015. We assume that the baseline washing machine corresponds to the China Energy Label Level 5, and that washing machine consumes 0.464 kWh per day, or 169.38 kWh per year. The average lifetime of washing machines is assumed to be 12 years.

Fluorescent Ballasts

We assume that fluorescent tube lights account for 20% of the lighting fixtures in residences and 75% of lighting energy in the commercial sector in China. The baseline fluorescent ballast is estimated at 38.6W with a high-efficiency option using 33.6W. We assume fluorescent ballasts consume 56.38 kWh per year and assume commercial fluorescent ballasts consume 112.76 kWh per year. The lifetimes of fluorescent ballasts are assumed to be 15 years.

Water Heaters

Both electric and gas water heaters are used widely in Chinese homes, and solar water heaters are becoming popular. The capacity of typical electric and gas water heaters are about 11 liters per minute. We assume that the baseline electric water heater corresponds to the China Energy Label Level 5, and that electric water heater consumes 1.69 kWh per day, or 617 kWh per year. We assume that the baseline gas water heater corresponds to the China Energy Label Level 3, and that gas water heater consumes 0.0137 GJ per day, or 4.97 GJ per year. The lifetimes are both are assumed to be 15 years. We consider replacement of water heaters by high efficiency units using the same fuel, but also replacement with solar water heating. Solar water heaters in China include a small electric back-up heater, which we assume to be used to heat water 30% of the time.

Standby Power

Standby power consumption is a feature of a wide range of products, including major appliances, consumer electronics and home entertainment equipment. This mode of power consumption is increasingly shown to be a major source of energy demand, and has become a prominent candidate for efficiency improvement (IEA, 2001). Reduction of standby power is typically very inexpensive to achieve through redesign of electronic components. The Chinese government has not moved to bring standby power into a single regulatory framework for all products, but it does include the reduction of standby power as part of individual rulemakings governing active mode use for each product type. For simplicity, we include standby power as an aggregate category, and assume an 80% reduction in standby power per product, from 5W to 1 W, at a cost of two cents per kWh. We assume that the average product using standby power lasts 8 years.

Electric Induction Ranges

Electric induction ranges are very popular in China. This technology uses an inductive core and high frequency current to produce heat, and has a much higher efficiency than traditional electric stoves. The key technology of higher efficiency products is the reduction of magnet leakage. We assume that the baseline electric induction range corresponds to the China Energy Label Level 5, and that electric induction range consumes 1.19 kWh per day, or 437.33 kWh per year. The lifetime of induction ranges is assumed to be 15 years.

Residential Boilers

An increasing number of Chinese homes use boilers for space heating. Residential boilers typically use natural gas for fuel, with diesel used less often. Residential boilers space heating equipment are subject to the same national standard as gas water heaters. We assume that the baseline gas residential boiler corresponds to the China Energy Label Level 3, and that gas residential boiler consumes 14.73 GJ per year[8] and has a lifetime is 17 years.

Commercial Boilers

Historically, Chinese commercial boilers used coal but increasingly, they are fired with natural gas, with a small fraction using diesel or methanol. We assume that the baseline commercial boiler corresponds to the China Energy Label Level 3. We estimate annual energy consumption assuming that 70kW unit operates for 4 hours per day for 6 months, resulting in gas consumption of 181.4 GJ per year. We assume a lifetime of 20 years for commercial boilers.

Industrial Motors

We assume that the baseline industrial motor corresponds to the China Energy Label Level 3. Annual energy consumption for motors is based on an assumption of 3000 hours per year. With this assumption, the baseline 10 HP industrial motor consumes 22500 kWh per year, 50 HP motors consume 112500 kWh per year and 100HP industrial motors consume 225000 kWh per year. The distribution of motor capacities in China is assumed to be in a 10:5:1 ratio for 10 HP, 50HP and 100 HP, respectively. The lifetime for all motors is assumed to be 12 years.

Distribution Transformers

Distribution Transformers are designed to reduce the voltage of the electricity coming from the electric grid, to a lower voltage which is applicable for appliances and other electricity-consuming systems. There is a national standard to regulate distribution transformers in China, with "minimum allowable values of energy efficiency and the evaluating values of energy conservation for three-phase distribution transformers" published in 2006. The typical baseline efficiency of distribution transformers is 95% (at 80% load). While there is not a Chinese labeling system for distribution transformers, an industry-recognized standard for high-efficiency transformers exists at 98.1% to 99.0% (at 80% load), depending on transformer capacity. Using these efficiency values, we calculate electricity losses from 5 size categories of transformers assuming that transformers are loaded 25% of the time. Distribution transformer lifetime is assumed to be 30 years.

3.2. Cost of Conserved Energy Calculation

As mentioned in the previous section, annual unit energy consumption (*UEC*) and equipment price (*Price*) are shown for all product classes considered in the analysis in the Appendix, unless otherwise specified[9]. These parameters are used in the calculation of cost of conserved energy according to Equation 1 by comparing each design option to the baseline, according to:

$$I = Price_{DesignOption} - Price_{Baseline}$$

and

$$S = UEC_{Baseline} - UEC_{DesignOption}$$

Table 1. Summary of Appliance Definitions and Retail Price Source

Appliance	Baseline Definition	High Efficiency Definition	Retail Price Source	Number of Observations
Incandescent Lamps	60 Watt Incandescent Lamp	15 Watt CFL	Haolives.com	282
Refrigerators	China Energy Label - Level 3	China Energy Label - Level 1	price.ea3w.com	830
Residential AC (Stable Speed)	China Energy Label - Level 5	China Energy Label - Level 2	detail.zol.com.cn	3800
Residential AC (Variable Speed)	China Energy Label - Level 5	China Energy Label - Level 2	detail.zol.com.cn	731
LCD TV	China Energy Label - Level 3	China Energy Label - Level 1	price.ea3w.com	480
PDP TV	China Energy Label - Level 3	China Energy Label - Level 1	price.ea3w.com	332
Washing Machines (Vertical)	China Energy Label - Level 5	China Energy Label - Level 1	price.ea3w.com	60
Washing Machines (Front Loading)	China Energy Label - Level 5	China Energy Label - Level 1	price.ea3w.com	60
Fluorescent Ballasts (Res.)	Magnetic	Electronic	Zhigou.com	10
Electric Water Heaters	China Energy Label - Level 5	China Energy Label - Level 1	price.ea3w.com	492
Induction Ranges	China Energy Label - Level 5	China Energy Label - Level 1	price.ea3w.com	88
Gas Water Heaters	China Energy Label - Level 3	China Energy Label - Level 1	price.ea3w.com	38
Gas Boilers (Residential)	China Energy Label - Level 3	China Energy Label - Level 1	163.com	45

Appliance	Baseline Definition	High Efficiency Definition	Retail Price Source	Number of Observations
Commercial Lighting	Magnetic	Electronic	Zhigou.com	10
Commercial Gas Boilers (Space Heating)	China Energy Label - Level 3	China Energy Label - Level 1	detail.china.alibaba.com	428
Commercial Room AC	China Energy Label - Level 5	China Energy Label - Level 2	detail.china.alibaba.com	3800
Industrial Motors	China Energy Label - Level 3	China Energy Label - Level 1	detail.china.alibaba.com	239
Distribution Transformers	95% Efficiency	98.1% - 99% Efficiency	detail.china.alibaba.com	305

The parameters used in calculation of q in Equation 2 are as follows:

Product Lifetime (L) – Average number of years that a product is used before failure and retirement. Lifetimes vary by product class and are estimated from manufacturer reports, or from survey data.

Discount Rates – In order to evaluate cost-effectiveness to consumers, the analysis takes into account the real cost of financing for Chinese consumers. As a reference, we considered interest rates offered by the People's Bank of China as representative of consumer financing costs. Interest rates for consumers at the time of the study were 5.6%[10]. Discount rates for the commercial and industrial sector were assumed to be somewhat higher, at 6.6% and 7.7% respectively.

Energy Prices – Current consumer electricity and natural gas prices do not fully reflect the cost of production and include cross-subsidization between tariff groups. For example, residential electricity prices range by region from about 0.35 RMB/kWh (0.05 USD/kWh) to 1.1 RMB (.17 USD/kWh)[11]. Current rates in Beijing are 0.48 RMB/kWh (0.073 USD/kWh) for residential customers 0.64 RMB/kWh (0.097 USD/kWh) for commercial customers and 0.57 RMB/kWh (0.086 USD/kWh) for small industrial consumers[12]. Aggressive moves by the Chinese government to raise consumer electricity prices are expected to continue, though with timing sensitive to the state of consumer price inflation. We therefore assume that by 2015 electricity rates for all consumer types will double. Electricity price assumptions are shown in Table 2.

Table 2. 2015 Electricity Price Assumptions

Sector	2010 RMB/kWh	2010 USD/kWh	2015 RMB/kWh	2015 USD/kWh
Residential	0.48	0.073	0.96	0.15
Commercial	0.64	0.097	1.28	0.19
Small Industry	0.57	0.086	1.14	0.17

Table 3. 2015 Natural Gas Price Assumptions

City	Residential Use	Industry Use
Beijing	2.05	2.35
Shanghai	2.5	3.3
Tianjin	2.2	2.4
Chongqing	1.4	1.67
Nanjing	2.2	2.35
Hangzhou	2.4	2.5
Chengdu	1.43	1.66
Wuhan	2.3	2.43
Urumqi	1.36	1.85
Guangzhou	3.45	3.95
Shenzhen	3.5	4.8
Average		
RMB/m^3	2.25	2.66
USD/1000m^3	341	403
USD/MJ	0.009	0.010
USD/GJ	8.8	10.4
2015 Prices	13.2	15.5

Source: China Natural Gas Report 2009 - C12009720CN.

Similarly, natural gas prices have shown a rapid increase in recent years. In fact, price trends indicate a doubling of retail natural gas prices, from 700 RMB per m^3 in 2001 to 1400 RMB per m3 in 2009[13]. We assume that natural gas prices will increase an additional 50% by 2015. We apply industrial natural gas prices to the commercial sector. Assumptions for natural gas prices are shown in Table 3.

Using these parameters, we calculate cost of conserved energy for each design option for each product class. The results of this calculation, shown in the Appendix tables, are the basis of construction of the efficiency scenario.

As stated above, the target efficiency level chosen is that which *maximizes efficiency while providing a net benefit to consumers*. Following this definition, we identify the target UEC for each product class as the lowest UEC for which cost of conserved energy is below the utility price.

To illustrate the construction of the efficiency scenario, we consider the example of room air conditioners. The top portion of Table 4 shows *UEC*, *Price* and cost of conserved energy (*CCE)* for each design option for room air conditioners. Room air conditioners (RAC) are used in both residential and commercial settings. In the residential sector, about 70% of the RAC market in China are heat pumps (reversible mode) and are therefore used for both heating and cooling. In the commercial sector, we assume that heating is generally supplied from other sources (such as district heat or boilers), and therefore RAC is used for cooling only in this sector. The revised efficiency standard that went into effect in July 2010 set Level 3 as the mandatory minimum; therefore we take this level as the baseline, and consider Level 1 as the high-efficiency level. In order to calculate cooling and heating energy, we assume that the typical RAC unit is 1.125 kW and that it operates 4 hours per day every day during a 3 month cooling period. In commercial applications, we assume that the air conditioner is used for 8 hours per day in the same cooling season. The resulting annual baseline energy consumption estimates are 385 kWh per year for residential cooling and 770 kWh per year for commercial cooling.

Where the same unit is used for heating (residential sector only), we make the following regional assumptions:

- Northern Region: 20% of heat pump market - 6 hours per day for two weeks (shoulder period of heating season when district heating is turned off).
- Transition Region: 80% of the heat pump market - 6 hours per day for 3 months.

These assumptions result in a weighted average baseline consumption of 891 kWh per year for heating and cooling.

Cost of conserved energy is calculated from Equations 1 and 2 using UEC, retail price, and the assumptions of product lifetime and appropriate discount rates. The result is a cost of conserved energy of 0.44 USD for residential cooling only, 0.095 USD for residential heating plus cooling and 0.116 USD for commercial cooling. The result is that the current Level1 is not a cost effective substitution to the current Level 3 baseline for consumers using cooling only, but it is cost-effective for heat pump users. In order to evaluate the total cost effective savings potential from this end use, we take the weighted average, assuming an improvement from 891 kWh to 669 kWh for heat pump (HP) users, and no improvement for AC-only users. The weighted average improvement is from 739 kWh to 584 kWh, an improvement of 21%. In the commercial sector, since the hours of operation are twice as high, replacement of Level 3 room air conditioners with Level 1 units is found to be cost effective, with a cost of conserved energy of 0.116 USD. The corresponding energy consumption reduction from 770 kWh to 578 kWh constitutes a 25% improvement.

Table 4. Cost of Conserved Energy Calculation for Room Air Conditioners

	Room AC								
Class	Residential AC (Cooling)			Residential HP (Cooling + Heating)			Commercial RAC		
Market share	30%			70%			100%		
Lifetime (years)	12			12			12		
Q	0.116			0.116			0.123		
Baseline (Level 3)	UEC (kWh)	Price ($)	CCE	UEC (kWh)	Price ($)	CCE	UEC (kWh)	Price ($)	CCE
	385	367		891	367		770	367	
High Efficiency (Level 1)	289	548	0.44	669	548	0.095	578	548	0.116
In-Class Target UEC	385			669			578		
Target End-Use UEC	584						578		
End-Use Baseline UEC	739						770		
Target End-Use CCE	0.095						0.116		

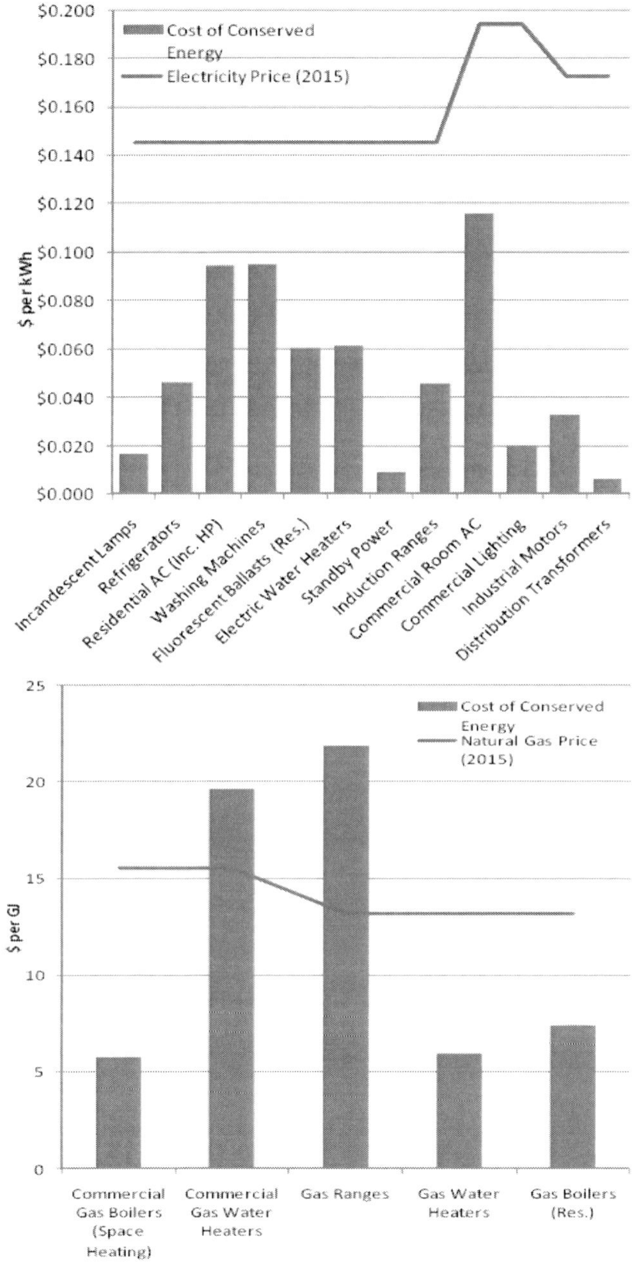

Figure 2. Cost of Conserved Energy and Energy Prices.

The Appendix tables show the calculation of CCE for each product class, and calculate weighted average target UEC and CCE for each appliance group. The cost of conserved energy for all appliance groups is compared to utility prices in Figure 2.

The main inputs to the construction of the two scenarios, the *Base Case* and the *Business Case* scenario are the baseline *UEC* and the UEC established by CCE in Figure 2. We call this the *Business Case UEC*. Baseline UEC, *Business Case* UEC, Percent Improvement and Cost of Conserved Energy are presented in Table 5.

Table 5. Unit Level Efficiency Improvement and Cost of Conserved Energy

Equipment	Baseline UEC	Business Case UEC	Percent Improvement	Cost of Conserved Energy
Electric Equipment	kWh	kWh	%	USD/kWh
Standby Power*	17.2	3.4	80%	0.009
Incandescent Lamps	50	13	75%	0.016
Industrial Motors	63281	61964	2.1%	0.020
Distribution Transformers	11176	2639	76%	0.006
Induction Ranges	437	398	9%	0.045
Refrigerators	548	292	47%	0.046
Fluorescent Ballasts (Res.)	56	49	13%	0.060
Electric Water Heaters	617	185	70%	0.061
Commercial Lighting	113	98	13%	0.020
Residential AC (Inc. HP)	739	584	21%	0.095
Washing Machines	169	92	46%	0.095
Commercial Room AC	770	578	25%	0.116
Fuel Equipment	GJ	GJ	%	USD/GJ
Commercial Gas Boilers (Space Heating)	181	159	13%	5.79
Gas Water Heaters	4.97	1.49	70%	5.92
Gas Boilers (Res.)	14.7	12.9	13%	7.37

* Per device using standby power (5W per device).

Table 5 shows that energy efficiency improvements can be made to a wide variety of equipment that will provide not only energy savings, but financial benefits to consumers. It also demonstrates the importance of performing this type of analysis at the appliance group level, since the cost-effective potential varies widely between appliance groups.

4. National Level Energy Savings Opportunities

Because of the modular structure of the BUENAS model (see Figure 1), once the inputs are established it is a relatively straightforward process to construct the two energy demand scenarios and compare them to calculate savings potential. The full details of the calculation of energy demand are provided in (McNeil 2008) and are omitted here.

4.1. Energy Savings and Emissions Reductions

Site energy savings is the basis for all national impacts calculations. Site energy demand refers to electricity and natural gas consumed in a home or business, and does not include fuel inputs in generation of electricity, or losses in transmission or distribution. Site energy is the energy affected most immediately by efficiency improvement. It is also the energy consumption that appears on consumer utility bills, and forms the basis for the cost-benefit analysis detailed above.

Site energy consumption is calculated by BUENAS for both the *Base Case* and *Business Case* scenarios. Energy *activity* is the same in both cases[14], so the difference between them is driven by the trend in *marginal intensity*, that is, the UEC of products sold in each year. The UEC for the two scenarios are identical until the policy implementation date of 2015[15]. After that date, the efficiency target in the *Business Case* is the high efficiency level determined by cost-benefit analysis, while it remains at the baseline efficiency level in the *Base Case*. The difference in UEC in the two scenarios applies only to new products – in this way, the policy modeled has the structure of a minimum efficiency performance standard, and does not imply retrofits of existing equipment. By 2016 overall energy demand of stock in the *Business Case* is only slightly lower than the *Base Case*, because only one year's sales are affected by the policy. Moving through the forecast, LEAP tracks the gradual flow of high efficiency products into the stock and the retirement of less efficient ones, so that the average stock UEC gets closer to the target level. Depending on the lifetime of the product, the entire stock may not be converted by 2030, since some low-efficiency products installed before 2015 will survive. Figure 3 shows the evolution of site energy savings by appliance group. From 2015 onward, energy savings grows for all products as high efficiency products begin to penetrate the stock in the *Business Case*.

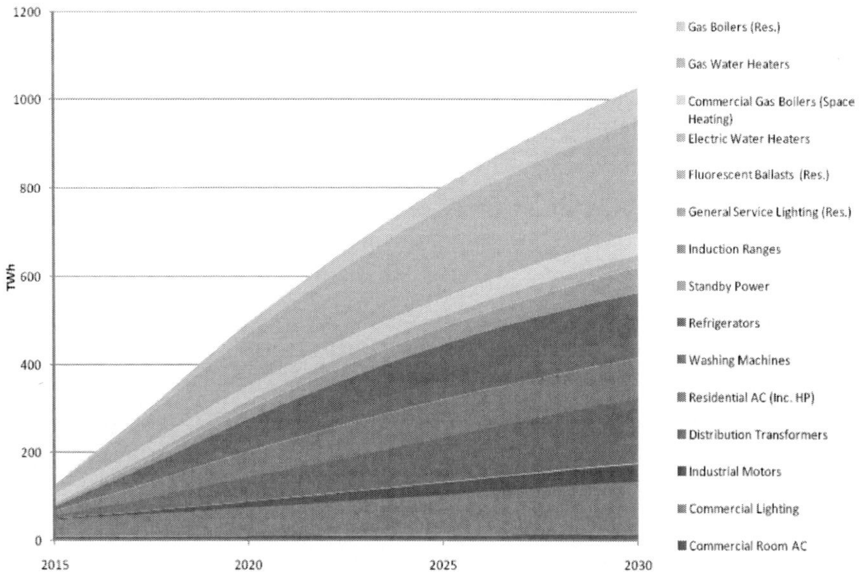

Figure 3. Site Energy Savings by Appliance Group– 2015-2030.

Site energy savings results are summarized in Table 6. Total savings for all appliance groups totals 1028 TWh in the year 2030. Cumulative savings through 2030 total 10,000 TWh.

Emissions reductions are calculated directly from energy savings according to a carbon factor. The carbon factor for electricity includes fuel inputs to generation, and accounts for transmission and distributions losses. The carbon factor taken from (Zhou 2011) is 0.98 kg CO_2/kWh in 2015, decreasing to 0.96 kg CO_2/kWh in 2030. Carbon factors for natural gas and fuel oil are assumed to remain constant at 0.202 and 0.264 kg/CO_2, respectively.

Emissions reductions from energy savings determined by multiplying energy savings by carbon factors are shown in Table 6. Total mitigation in the *Business Case* is found to be 717 mt CO_2 in 2030 and 7234 mt CO_2 over the entire forecast. Figure 4 shows the contribution to cumulative CO_2 mitigation from all appliance groups. In addition, we also evaluate the amount of other pollutants avoided by energy savings, including SO_2[16], NO[17] and mercury (Hg)[18].

Several conclusions can be drawn from Table 6 and Figure 4. First, emission reduction potential is well distributed among end uses and sectors. The largest potential exists for commercial lighting, residential air conditioning and distribution transformers, each of which could provide over 100 TWh of electricity demand reduction in 2030 and about 1300 mt CO2 over the forecast period. Savings in commercial lighting is possible due to the prevalence of magnetic ballasts for fluorescent lamps, which can be replaced with electronic ballasts.

Air conditioning savings includes reduction of heating energy from reversible models (heat pumps). Finally, distribution transformers show significant savings even taking into account the recent standards rulemaking because of the scale of distribution losses at the national level, even though total efficiency improvement is small on a percentage basis. Much of the remaining mitigation potential comes from major appliances such as refrigerators and washing machines. Significant savings can be gained from replacement of gas and electric water heaters with solar water heating systems, which already have a significant market share in China.

Table 6. Energy Savings and Pollutant Mitigation by Appliance Grou

Appliance Group	Final Energy Savings			Emissions Mitigation					
	In 2020	In 2030	Through 2030	In 2020	In 2030	Through 2030			
	TWh			mt CO_2		mt CO_2	mt SO_2	mt NO	t Hg
Gas Water Heaters	118	255	2448	24	52	494	6.2	8.1	403.9
Distribution Transformers	57	148	1273	60	146	1294	3.2	4.2	210.1
Commercial Lighting	63	121	1249	66	119	1277	3.2	4.1	206.1
Residential AC (Inc. HP)	57	90	1020	60	89	1041	2.6	3.4	168.3
Refrigerators	49	100	991	51	98	1009	2.5	3.3	163.5
Gas Boilers (Res.)	26	75	607	5	15	123	1.5	2.0	100.1
Commercial Gas Boilers (Space Heating)	32	49	567	7	10	114	1.4	1.9	93.5
Washing Machines	25	50	505	27	49	514	1.3	1.7	83.3
Industrial Motors	14	40	338	15	40	343	0.9	1.1	55.8
Incandescent Lamps	10	38	282	11	38	286	0.7	0.9	46.5
Standby Power	12	20	210	12	19	214	0.5	0.7	34.6
Electric Water Heaters	12	17	203	12	17	207	0.5	0.7	33.5
Commercial Room AC	11	14	185	11	14	190	0.5	0.6	30.5
Fluorescent Ballasts (Res.)	7	9	108	7	9	110	0.3	0.4	17.7
Induction Ranges	1	2	16	1	2	16	0.0	0.1	2.6
Total	494	1028	10001	368	717	7234	26	33	1650

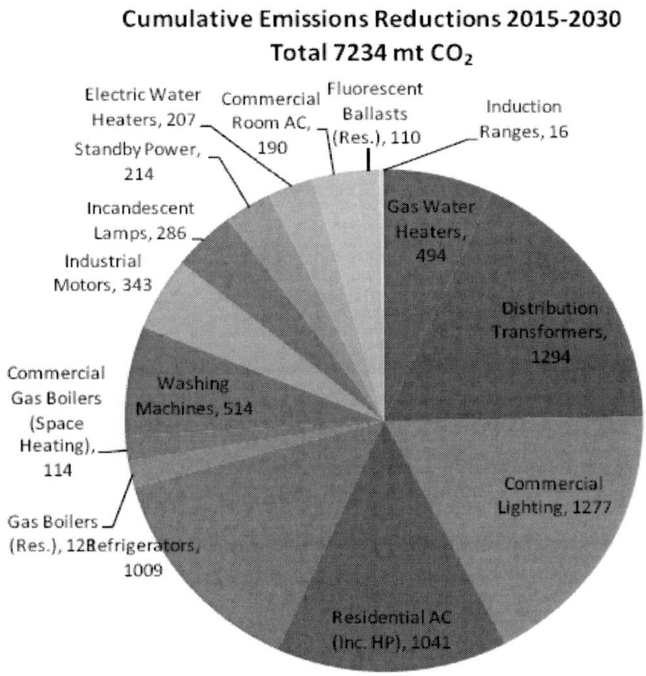

Figure 4. Cumulative CO2 Emissions Reductions 2010-2030.

4.2. Consumer Financial Impacts

By construction, the *Business Case* implements energy efficiency in a way that is cost-effective to consumers. Because this study insisted on quantifying investments needed to improve efficiency relative to the base case technology, the necessary information to evaluate these investments and financial benefits of energy savings, and therefore net financial impacts to consumers, is available for all appliance groups considered.

Recalling the definition of cost of conserved energy from Equation 1:

$$CCE = \frac{I \times q}{S}$$

The denominator of this equation I x q is the annualized equipment investment necessary to yield an annual energy savings S. BUENAS calculates the total savings $ST(y)$ in each year, given by:

$$S_T(y) = S \times Stock'(y)$$

In this equation, *Stock'(y)* is the affected stock, that is, the number of units operating in the stock that were installed after the policy implementation date, and are each providing a savings *S* relative to the *Base Case*. Likewise, the total annualized investment in each year *IT(y)* x *q* is given by:

$$I_T(y) \times q = I \times q \times Stock'(y)$$

Substituting Equation 1, and cancelling terms, yields:

$$I_T(y) \times q = S_T(y) \times CCE$$

In other words, total annualized investment can be calculated for each appliance group by multiplying its total energy savings by the cost of conserved energy shown in Table 5.

Financial savings from energy savings is given simply by the utility price in each year multiplied by the total energy savings *ST(y)*. Net financial impacts are then given by:

$$N(y) = S_T(y) \times (Utility\ Price - CCE)$$

Costs, Savings and Net Impacts calculated in this way are shown in Table 5. In evaluating the financial value of efficiency or other government programs, it is customary to take account of deferred benefits through a discount rate calculation. The resulting *Net Present Value* (NPV) of benefits is given by:

$$NPV = \sum_{y=2010}^{2030} \frac{N(y)}{(1 + DR)^{y-2010}}$$

In this equation, *DR* is a “societal” discount rate that parameterizes the preference for immediate returns on public investments. We consider two scenarios in which the societal discount rate is taken to be 3% or 7%. Cumulative equipment costs, energy bill savings, net savings and *NPV* are shown in Table 5. The table shows positive net savings for all appliance groups, which is not surprising, since the target efficiency levels were constructed to be cost-effective. The table shows that cost-effective efficiency improvements require an investment of 354 billion USD over the next 20 years, but these investments will return over three times as much over the same period, for a net savings of 846 billion USD, or of order of six hunred USD per capita. The net present value is 554 billion USD assuming a discount rate of 3%, and 328 billion with a 7% discount rate. Of the appliance groups studied, residential air conditioners require the largest investment at 97 billion USD, but also have the one of the highest payoffs at 148 billion USD. The highest payoffs are found in the commercial sector, with 212 billion USD for distribution transformers and 217 billion for commercial lighting, which is driven by implementation of electromagnetic ballasts. Both of these technologies are highly cost effective, with net savings of over 200 billion USD.

Table 7. Cumulative Financial Impacts of Efficiency Improvement

Appliance Group	Cumulative Financial Impacts				
	Cost	Savings	Net Savings	NPV @ 3% DR	NPV @ 7% DR
	Billions USD				
Commercial Lighting	25.0	242	217	145	88.0
Distribution Transformers	7.7	220	212	137	79.7
Refrigerators	45.6	144	98.5	64.3	37.8
Gas Water Heaters	52.1	116	63.8	41.5	24.4
Residential AC (Inc. HP)	96.6	148	51.8	34.2	20.4
Industrial Motors	11.0	58.4	47.4	30.5	17.6
Incandescent Lamps	4.6	41.0	36.4	23.2	13.2
Standby Power	1.9	30.5	28.6	18.9	11.3
Washing Machines	47.9	73.4	25.5	16.7	9.8
Commercial Gas Boilers (Space Heating)	11.8	31.7	19.9	13.4	8.3
Commercial Room AC	21.4	35.9	14.5	9.9	6.2
Electric Water Heaters	12.4	29.6	17.1	11.3	6.8
Gas Boilers (Res.)	16.1	28.7	12.6	8.1	4.7
Fluorescent Ballasts (Res.)	6.4	15.6	9.2	6.1	3.7
Induction Ranges	0.7	2.3	1.6	1.0	0.6
Total	354	1200	846	554	328

Finally, financial impacts, emissions savings and their relationship can be shown using a "conservation supply curve". This unique way of expressing the cost and benefits of carbon mitigation measures has become very widespread in the literature because of the key information it conveys. A conservation supply curve for the *Business Case* is presented in Figure 6. The x-axis shows cumulative carbon mitigation and expresses the relative importance of each appliance group. The total extent of the curve is 7234 mt CO2, as shown in Table 6. The y-axis displays relative affordability according to cost of conserved energy. The blocks corresponding to each measure are ordered with increasing cost of conserved energy, from left to right.

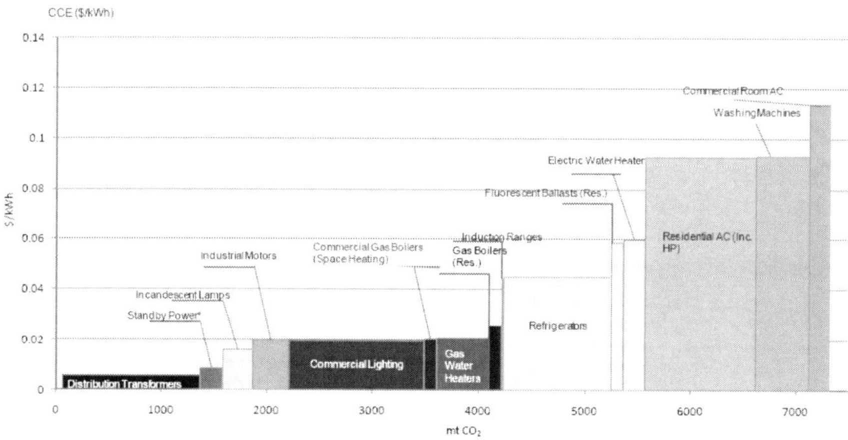

Figure 5. Financial Impacts for All End Uses 2010-2030.

Finally, we note that there are other benefits to the energy savings achieved in the *Business Case* besides the direct energy and financial benefits. The effect of reduction of greenhouse gas emissions and resulting avoided costs are difficult to quantify, but could be very large. One metric to consider the order of magnitude of the value of these types of impacts is the assumption of a carbon price. The assumption of a price of 25 USD per ton of carbon dioxide yields an additional 205 billion USD of savings, while a 100 USD per ton price yields 822 billion additional USD, nearly doubling the total.

The negative impacts of emissions of SO2 and NO from power plants are well-known (see, for example, EPA 2010), including acid rain, acidification of watersheds and lakes, and respiratory illness from inhaling particulates. Likewise, the reduction of mercury emissions from coal-burning power plants reduces fish contamination, which is now recognized as a major health risk.

We do not try to quantify the health impacts of reduction of these emissions, only point out the obvious – that savings due to efficiency is equivalent to installation of clean electricity generation. In the *Business Case*, this reduction provides a large net financial benefit to consumers, which may not be true with alternatives.

CONCLUSION

The *Business Case* analysis found additional potential for cost-effective efficiency improvement in China for seventeen appliance groups in the residential and commercial building sectors and industrial motors. Efficiency improvement for these technologies could deliver three times as much financial benefit to Chinese households and business than the investment needed to implement them. In addition to direct financial benefits, impacts on greenhouse gas emissions and are significant. Total net impacts from additional deployment of high efficiency technology include:

Energy savings:

- 490 billion kWh per year in 2020
- 1015 billion kWh per year in 2030
- A total of 10,000 billion kWh cumulatively through 2030

Cumulative greenhouse gas emissions mitigation:

- 7730 million metric tons of CO2 through 2030
- 26 million metric tons of SO2 through 2030
- 33 million metric tons of NO through 2030
- 1650 metric tons of mercury through 2030

Financial impacts to consumers through 2030:

- Equipment investment of 350 billion USD
- Energy bill savings of 1200 billion USD
- Net savings of 846 billion USD

The “business case” analysis shows that the Chinese market already has access to efficiency technologies that could provide Chinese consumers with a financial benefit and make a dent in the growth of Chinese emissions if widely adopted. Most of the equipment studied has been the subject of at least one efficiency standard, but opportunities for improvement are not exhausted. To some extent, therefore, the savings potential estimated by this study can be captured through expansion and aggressive pursuit of existing Chinese government policies. It should also be noted that many of the technologies included in the "business case" scenario were not available ten to twenty years ago, or at least weren't be shown to be cost effective. These technologies have become available and cost-effective through research, new materials and components, improvements in production processes, or changes in design of systems. Likewise, we expect that a similar analysis performed 10 years from now will show improvements not accessible to the current study due either to lack of data or prohibitively high cost of "prototype" technologies.

Because the rigor of the methodology used to evaluate cost-effectiveness requires a significant amount of technical data, we only cover a subset of equipment types for which significant savings potential might be available. In particular, the appliance groups covered are limited to buildings applications. For this sector, however, we believe a large fraction of energy demand is accounted for. For this reason, while the overall savings potential is large, it cannot be interpreted as "comprehensive".

Finally, we believe this study to be among the few to attempt to evaluate the "economic" potential of efficiency improvement in China in a transparent way. In addition to demonstrating significant savings potential, we hope that it demonstrates a clear and consistent methodology for creation and expansion of alternative energy scenarios in China. Additional scenarios that could be explored include the potential impact of carbon taxes, cap-and-trade, R&D investments and other policy- or market-based drivers. The ability of the research community to utilize this type of analysis to inform government and private sector decision makers will depend largely on investments made in development of the type of data used here, both more widely and with greater frequency.

REFERENCES

Aden, N., Qin, Y., and Fridley, D., 2010. *Lifecycle Assessment of Beijing-Area Building Energy Use and Emissions: Summary Findings and Policy Applications.* LBNL, Berkeley, CA. http://china.lbl.gov/sites/china.lbl.gov/files/LBNL-3939E.pdf

China Daily. 2010. "China to subsidize 150m energy-efficient bulbs" June 30, 2010

Cheung, R. and Kang, A., 2008. *China's Booming Energy Efficiency Industry.* World Resources Institute (WRI). *http://pdf.wri.org/chinas_booming_energy_efficiency_industry.pdf*

CNIS, 2010. *White Paper: The Status of Energy Efficiency Developments in Energy-using Products in China.* Beijing: Standards Press of China.

De la Rue du Can, S., Letschert, V., McNeil, M., Zhou, N., and Sathaye, J., 2009. *Residential and Transport Energy Use in India: Past Trend and Future Outlook.* LBNL, Berkeley, CA. *http://ies.lbl.gov/drupal.files/ies.lbl.gov.sandbox/LBNL-1753E.pdf*

Delio, E.A., Lall, S., and Singh, C., 2009, *Powering Up: The Investment Potential of Energy Service Companies in India.* World Resources Institute. http://pdf.wri.org/powering_up_full_report.pdf

EIA, 2010. *Annual Energy Outlook 2010*, U.S. Department of Energy - Energy Information Administration.

Energy Foundation (EF), 2006. "*China Electric Utilities Program: Strategy and Project Descriptions".* http://www.efchina.org/FReports.do?act=detail&id=158

EPA, 2010. U.S. Environmental Protection Agency, *2009 Annual Report: Energy Star and Other Climate Protection Partnerships.* December 2010. http://www.energystar.gov/ia/partners/publications/pubdocs/2009%20CPPD%20Annual%20 Report.pdf

Feng, Xinbin,2010 "Interview with Feng Xinbin - If Energy Efficient Light Bulb Breaks... ". Liaoning Daily. December 14, 2010. http://epaper.lnd.com.cn/lnrb/20101214/index.htm

Finamore B., M. Yew and B. Ho., 2010. *Taking Action to Meet its Climate Pledge - China Enacts National Energy Efficiency DSM Regulations to Dramatically Scale Up Investments in Energy Efficiency* http://switchboard.nrdc.org/blogs/bfinamore/taking action to meet its clim.html

Fridley D., N. Aden, N. Zhou, J. Lin, 2007. *Impacts of China's Current Appliance Standards and Labeling Program to 2020*, March 2007. http://www.clasponline.org/files/Task%201%20FINAL%20Report-%20METI.pdf

Fridley, D., Zheng, N., Zhou, N., Aden, N., Lin, J., Chen, J., and Sakamoto, T. 2008. *China Refrigerator Information Label: Specification Development and Potential Impact.* LBNL246E, LBNL, Berkeley, CA.

*http://china.lbl.gov/sites/china.lbl.gov/files/LBNL_246E._China_Refrigerator_Information_* Label. Feb2008.pdf

Hu Zhaoguang, D. Moskovitz and Z Jianping. *Restructured Power Industry: How Regulation and Policy Can Deliver Demand-Side Management Benefits to a Growing Economy and a Changing Power System*, Energy Sector Management Assistance Program (ESMAP), December 2005, http://www.raponline.org/Pubs/China/Dec05ChinaDSM.pdf

IEA, 2001. International Energy Agency. *Things that go Blip in the Night – Standby Power and how to Limit It*. OECD. 2001

IEA, 2009, Gadgets *and* Gigawatts -- *Policies for Energy Efficient Electronics*, Paris. http://www.iea.org/textbase/nppdf/free/2009/gigawatts2009.pdf

Letschert, E., M.A. McNeil and N. Zhou, 2009. *Residential Electricity Demand in China –Can Efficiency Reverse the Growth?*. 5th International Conference on Energy Efficiency in Domestic Appliances and Lighting (EEDAL 09). Berlin, Germany. LBNL-2321E

Moskovitz, D., F. Weston, and J Lin, 2007, *Meeting China's Energy Efficiency and Environmental Goals with Efficiency Power Plants (EPPs)*. June 2007, Report of Regulatory Assistance Project

Moskovitz, D, 2005. *Meeting China's Energy Efficiency Goals Means China Needs to Start Building Efficiency Power Plants (EPP)*, November 2005. http://www.raponline.org/Pubs/China/ChinaDailyEPP.pdf

Price, L., Mark D. Levine, Nan Zhou, David Fridley, Nathaniel Aden, Hongyou Lu, Michael McNeil, Nina Zheng, Yining Qina and Ping Yowargana, 2011 "Assessment of China's Energy-Saving and Emission-Reduction - Accomplishments and Opportunities During the 11th Five Year Plan". Energy Policy 39(4) pp 2165-2178. LBNL-3385E

Shen,B, B. Finamore, and M. Yew, 2009, *Promoting Energy Efficiency as a Cost-Effective Resource in China: A Review of Jiangsu's Efficiency Power Plant Pilot*, ACEEE Proceedings of the 2009 Summer Study, http://www.aceee.org/sites/default/files/publications/proceedings/SS09_Panel4_Paper01.pdf

Wang, G., 2010. *Energy Efficiency Progress in China*. Presentation made at 35th EGEE&C Meeting, April 2010, Wellington, New Zealand

Zhou N., 2008. *Status of China's Energy Efficiency Standards and Labels for Appliances and International Collaboration*, March 2008. http://china.lbl.gov/sites/china.lbl.gov/files/LBNL 251E. Status of Chinas Energy Efficie ncy Standards and Lables. Mar2008.pdf

Zhou et. al., 2010, *Analysis and Potential Energy Saving and CO2 Emission Reduction of Home Appliances and Commercial Equipments in China*.

Proceedings of the 2010 ACEEE Summer Study on Energy Efficiency in Buildings

Zhou, N., Levine, M.D., and Price, L., 2010. *Overview of Current Energy Efficiency Policies in China* Energy Policy 38(11), November 2010, Pages 6439-6452

Zhou, N. 2011. Zhou, N., D. Fridley, M.A.McNeil, N.Zheng, J.Ke and M. Levine 2011. *China's Energy and Carbon Emissions Outlook to 2050.* Lawrence Berkeley National Laboratory Report 4472E.

APPENDIX – EFFICIENCY-COST RELATIONSHIP AND COST OF CONSERVED ENERGY CALCULATION FOR APPLIANCE GROUPS

Parameters used in Calculation of Cost of Conserved Energy:

Residential Consumer Discount Rate = 5.6%
Commercial Consumer Discount Rate = 6.6%
Industrial Consumer Discount Rate = 7.6%
Residential Electricity Price (2015) = $0.15 $/kWh
Commercial Electricity Price (2015) = $0.19 $/kWh
Commercial Electricity Price (2015) = $0.17 $/kWh
Residential Natural Gas Price (2015) $13.16 $/GJ
Commercial Natural Gas Price (2015) $15.53 $/GJ

Table A.1. Efficiency-Cost Relationship for Room Air Conditioners and Heat Pumps

	Room AC								
Class	Residential AC (Cooling)			Residential HP (Cooling + Heating)			Commercial RAC		
Market share	30%			70%			100%		
Lifetime (years)	12			12			12		
Q	0.116			0.116			0.123		
Level	UEC (kWh)	Price ($)	CCE	UEC (kWh)	Price ($)	CCE	UEC (kWh)	Price ($)	CCE
Baseline (Level 3)	385	367		891	367		770	367	
High Efficiency (Level 1)	289	548	0.44	669	548	0.095	578	548	0.116
In-Class Target UEC	385			669			578		

Class	Room AC		
	Residential AC (Cooling)	Residential HP (Cooling + Heating)	Commercial RAC
Target End-Use UEC	584		578
End-Use Baseline UEC	739		770
Target End-Use CCE	0.095		0.116

Table A.2. Efficiency-Cost Relationship for Water Heaters

Class	Water Heaters					
	Waterheaters (Electric)			Waterheaters (Gas)		
Market share	100%			100%		
Lifetime (years)	15			15		
Q	0.100			0.100		
Level	UEC (kWh)	Price	CCE	UEC (kWh)	Price	CCE
Baseline	617	176		5.0	235	
1	370	367	0.077	4.1	470	28.362
2	185	441	0.061	1.5	441	5.915
Target End-Use UEC	185			1		
End-Use Baseline UEC	617			5		
Target End-Use CCE	0.061			5.92		

Table A.3. Efficiency-Cost Relationship for Other Appliance Groups

Equipment	Lifetime	Q	Baseline UEC	Target UEC	Baselin e Price	Target Price	Target CCE
Electric Equipment	years	years	kWh	kWh	$	$	$/kWh
Refrigerator	15	0.100	548	292	323	441	0.046
Laundry (Washing Machine)	15	0.100	169	91.9	220	294	0.095
Cooking Products (Electric Induction Stoves)	15	0.100	437	398	34	51	0.045
Lamps (Incandescent)	5	0.235	50.4	12.6	1.47	4.11	0.016
End-Uses with more than one product class							
Fluorescent Lamp Ballast : Residential	15	0.100	56.4	49.0	0*	4.41	0.060
Fluorescent Lamp Ballast : Commercial	15	0.100	112.8	98.1	0*	4.41	0.030

Equipment	Lifetime	Q	Baseline UEC	Target UEC	Baseline Price	Target Price	Target CCE
Commercial Air Conditioning	12	0.123	48000	35692	17621	26432	0.088
Market Share of Residential = 50% Commercial = 50%				Baseline UEC	73.540	84.57	0.060
Equipment	Lifetime	Q	Baseline UEC	Target UEC	Baseline Price	Target Price	Target CCE
Electric Equipment	years	years	kWh	kWh	$	$	$/kWh
10 HP (62.5%)	10	0.146	22500	22032	176	294	0.037
50 HP (31.25%)	10	0.146	112500	110159	881	1322	0.027
100 HP (6.25%)	10	0.146	225000	220318	1028	2349	0.041
Distribution Transformers							
25 kVA (6%)	30	0.085	2738	1051	1028	1367	0.017
60 kVA (25%)	30	0.085	6570	1822	1175	1562	0.007
100 kVA (44%)	30	0.085	10950	2628	1909	2539	0.006
160 kVA (3%)	30	0.085	19710	3854	2496	3320	0.004
200 kVA (14%)	30	0.085	21900	4555	2937	3906	0.005
Fuel Equipment	years	years	GJ	GJ	$	$	$/GJ
Commercial Gas Boiler	20	0.091	181	2496	159	3935	5.79
Gas Boilers	17	0.092	14.7	734	13	881	7.37

End Notes

[1] For simplicity the high efficiency scenario assumes 100% of the market will reach the target level in 2015, a structure that closely resembles minimum efficiency performance standards. In the later years of the forecast, the scenario is not highly sensitive to the details of the market transformation.

[2] BUENAS is described completely in McNeil, M.A., V.E. Letschert and S.A.De la Rue du Can (2008). *Global Potential of Energy Efficiency Standards and Labeling Programs*. LBNL 760E.

[3] More information about the LEAP platform may be found at http://www.energycommunity.org

[4] This depends somewhat on the lifetime of the product. For refrigerators we may assume a 15 - year lifetime, but some refrigerators may last 20 years, so the turnover of the stock may not be complete by 2030.

[5] Other metrics such as life cycle cost and payback period establish cost effectiveness, but are not easily compared across disparate technologies and end uses.

[6] China National Development and Reform Commission "a phase-out of incandescent bulbs road map". (http://www.eeo.com.cn/Politics/beijing_news/2010/06/25/173754.shtml).

[7] http://life.yesky.com/113/11596113.shtml

[8] LBNL estimate – *China Energy 2050 Model*

[9] Additional products were evaluated for cost-effectiveness with the result that no cost-effective improvements were possible beyond the current standard. These were removed from further consideration, and are not included in the Appendix.

[10]http://www.global-rates.com/interest-rates/central-banks/central-bank-china/pbc-interest-rate.aspx

[11] Exchange rate of 6.6 RMB per dollar as of October 2010.

[12] http://www.bj.sgcc.com.cn/Contents/Channel_52/2009/1123/58255/content_58255.htm

[13] http://wenku.baidu.com/view/41c0e93610661ed9ad51f3ea.html

[14] It is possible to model, for example, the reduction of sales or fuel switching resulting from price increases associated with efficiency regulations. This effect is not captured in BUENAS.

[15] The exception is the phase-out of incandescent lamps, which begins in 2012 in the *Business Case*.

[16] SO2 Emissions rate of 2.55 g/kWh estimated from www.dianli.com

[17] NO Emissions rate of 3.3 g/kWh from (Zhou 2011)

[18] Mercury emissions rate of 0.165 mg/kWh from (Feng 2010)

In: Business Case Scenarios for Energy Efficiency ISBN: 978-1-62948-076-3
Editor: Sean C. Byrne

Chapter 3

Business Case for Energy Efficiency in Support of Climate Change Mitigation, Economic and Societal Benefits in India*

Michael A. McNeil, Jing Ke, Stephane de la Rue du Can, Virginie E. Letschert and James E. McMahon

Executive Summary

This study seeks to provide policymakers and other stakeholders with actionable information towards a road map for reducing energy consumption cost-effectively. We focus on individual end use equipment types (hereafter referred to as *appliance groups*) that might be the subject of policies - such as labels, energy performance standards, and incentives - to affect market transformation in the short term, and on high-efficiency technology options that are available today.

As the study title suggests, the high efficiency or *Business Case* scenario is constructed around a model of cost-effective efficiency improvement. Our analysis demonstrates that a significant reduction in energy consumption and emissions is achievable at *net negative cost*, that is, as a profitable investment

* This is an edited, reformatted and augmented version of Ernest Orlando Lawrence Berkeley National Laboratory, Publication No. LBNL-5344E, dated November 29, 2011.

for consumers savings are calculated assuming no additional costs to energy consumption such as carbon taxes. Savings relative to the base case as calculated in this way is often referred to as "economic savings potential".

The Indian energy demand picture is defined by very low average consumption that is growing very rapidly. For most end uses, this means that consumption in 2030 will be several times higher than it is today. Nearly half of Indian households are currently without electricity, and most do not own major appliances such as refrigerators and air conditioners. This situation is changing rapidly, however, with increased access to commercial energy sources and a swelling middle class with the purchase power to own and use major appliances and electronics. The low ownership/high growth situation means that the vast majority of equipment operating in 2030 will be installed after 2015. This means that efficiency measures implemented now will have maximal impact by the end of the forecast.

The organized commercial sector in India occupying large office buildings, retail outlets and hotels is still relatively small and consumes only a fraction of the energy as the household sector. Commercial building electricity consumption is growing, however, and savings there can be significant. Consumption is concentrated in lighting and air conditioning.

The industrial sector in India is a major energy consumer, with a large fraction of industrial electricity consumption passing through electric motors. Finally, losses in electricity distribution in India are high, partially due to a relatively inefficient stock of distribution transformers, which offer a significant opportunity for improvement.

Recognizing the need for enhanced energy efficiency in the face of exploding demand, the Government of India has already launched important programs to promote the adoption of high-efficiency equipment, including many actions already taken by the Bureau of Energy Efficiency (BEE) established as a direct consequence of the Energy Conservation Act of 2001. In a further important step, the Indian government announced its Action Plan on Climate Change (NAPCC) in June 2008, including creation of a National Mission on Enhanced Energy Efficiency. Finally, adoption of a proposed amendment to the Energy Conservation Act of 2001 would deepen the scope of government actions on energy efficiency.

So far, the Indian market has responded favorably to government efficiency initiatives, with Indian manufacturers producing a higher fraction of high-efficiency equipment than before program implementation. This study highlights both the financial benefit and the scope of potential impact for adopting this equipment, all of which is already readily available on the market.

Energy savings:

- 60 billion kWh per year in 2020
- 140 billion kWh per year in 2030
- A total of 1,250 billion kWh cumulatively through 2030

Cumulative greenhouse gas emissions mitigation:

- 1350 million metric tons of CO_2 through 2030

Financial impacts to consumers through 2030:

- Equipment investment of 38 billion USD
- Energy bill savings of 96 billion USD
- Net savings of 58 billion USD

The approach of the study is to assess the impact of short-term actions on long-term impacts. "Short-term" market transformation is assumed to occur by 2015, while "long-term" energy demand reduction impacts are assessed in 2030. In the intervening years, most but not all of the equipment studied will turn over completely. The 15-year time frame is significant for many products, in the sense that delay of implementation postpones economic benefits and mitigation of emissions of carbon dioxide. Such delays would result in putting in place energy-wasting technologies, postponing improvement until the end of their service life, or potentially resulting in expensive investment either in additional energy supplies or in early replacement to achieve future energy or emissions reduction targets.

The *Business Case* concentrates on technologies for which cost-effectiveness can be clearly demonstrated. The appliance groups studied are:

Residential End Uses
Incandescent Lamps
Refrigerators
Air Conditioners
Fluorescent Ballasts
Standby Power

Commercial and Industrial End Uses
Commercial Lighting
Commercial Air Conditioning
Industrial Motors
Distribution Transformers

Energy savings and greenhouse gas emissions mitigation for these appliance groups are summarized in Table ES-1.

Table ES-1. Energy Savings and Pollutant Mitigation by Appliance Grou

End Use	Final Energy Savings			Emissions Mitigation		
	In 2020	In 2030	Through 2030	In 2020	In 2030	Through 2030
	TWh			mt CO2		mt CO2
Air Conditioners	11.0	46.0	315	12.1	48.6	339
Standby	12.6	25.0	249	13.9	26.4	269
Incandescent Lamps	18.5	3.6	207	20.4	3.8	226
Refrigerators	6.2	23.7	172	6.8	25.1	185
Industrial Motors	3.8	12.0	95	4.2	12.7	103
Distribution Transformers	3.7	12.6	95	4.0	13.3	103
Commercial Lighting	2.2	8.0	59	2.4	8.4	63
Fluorescent Lamps	1.5	4.9	38	1.7	5.1	41
Commercial Cooling	0.8	3.2	22	0.8	3.3	24
Total	60	139	1251	66	147	1351

Since the study includes only appliance groups for which cost-effectiveness can be clearly demonstrated, the benefits determined represent only a subset of the economy-wide potential. Specifically, transportation end uses and industrial processes technologies are not covered, because data sufficient to include them were not possible to collect within the scope of the research. Likewise, the study does not include system approaches such as smart grids. These approaches to efficiency may have important impacts but the calculation of costs and benefits is not as straightforward as for individual pieces of equipment. In addition, the technologies analyzed represent a snapshot of what is currently on the market. Technological innovations are certain to occur over the coming decades, and these will likely present new opportunities for efficiency improvement, and exert downward pressure on costs.

Efficiency measures are determined to be cost-effective if the cost of conserved energy associated with them is less than the consumer's energy price, that is, the amount saved in energy bills is greater than the initial investment. The *Business Case* scenario is generated by identifying the *maximum efficiency improvement for which cost of conserved energy is lower than utility energy prices* (projected to 2015). The relative contribution to cumulative emissions for each appliance group is shown in Figure ES-1.

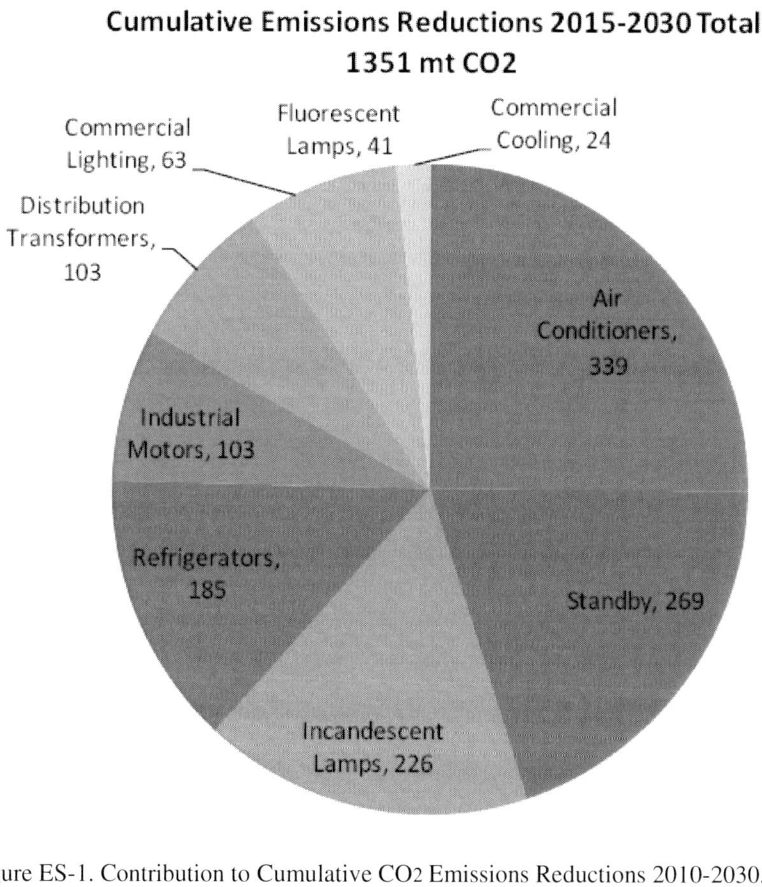

Figure ES-1. Contribution to Cumulative CO_2 Emissions Reductions 2010-2030.

Several conclusions can be drawn from Table ES-1 and Figure ES-1. First, emission reduction potential is well distributed among end uses and sectors. The largest potential exists for residential air conditioning, incandescent lamps and refrigerators and standby power, each of which could provide over 200 mt CO2 over the forecast period. The large savings from room air conditioners is driven by a combination of both the high cooling load in India and the related rapid growth of this end use in Indian homes. Savings from incandescent lamp phase out is high cumulatively, but low in terms of savings in 2030. This is because incandescent are assumed to be phased out by that year even in the Base Case – the Business Case for lamps is an acceleration of that phase out. Finally, significant savings potential is also shown for refrigerators, industrial motors and distribution transformers.

Table ES-2. Cumulative Financial Impacts of Efficiency Improvement

Appliance Group	Cumulative Financial Impacts				
	Cost	Savings	Net Savings	NPV @ 3% DR	NPV @ 7% DR
	Billions USD				
Incandescent Lamps	2.2	14.8	12.6	8.9	5.9
Standby	6.9	17.7	10.9	8.4	4.9
Air Conditioners	11.0	22.5	11.5	7.3	4.1
Refrigerators	4.9	12.2	7.4	4.7	2.7
Distribution Transformers	3.5	9.3	5.8	2.8	1.6
Commercial Lighting	0.7	5.4	4.6	1.6	0.9
Industrial Motors	4.9	9.3	4.4	1.2	0.7
Fluorescent Lamps	2.3	2.7	0.4	0.2	0.1
Commercial Cooling	1.6	2.0	0.4	1.1	0.6
Total	38	96	58	36	22

The analysis shows that cost-effective efficiency improvement could yield very significant financial benefits to Indian consumers. Table ES-2 shows positive net savings for all appliance groups, which is not surprising, since the target efficiency levels were constructed to be cost-effective. The table shows that cost-effective efficiency improvements require an investment of 38 billion USD over the next 20 years, but these investments will return over more than twice as much over the same period, for a net savings of 58 billion dollars, or of order of fifty dollars per capita. The present value of net savings is 36 billion USD assuming a discount rate of 3%, and 22 billion USD with a 7% discount rate.

Of the appliance groups studied, residential room air conditioners require the largest investment at 11 billion USD, but provide a payoff of 22.5 billion USD. Standby power generates similar savings, but with only half of the investment. Phasing out incandescent lamps is extremely cost effective compared to other appliance groups, with a payoff of over six times as high as the required investment.

1. INTRODUCTION

Some recent examples of studies that have identified potential energy savings from energy efficiency improvements include:

- Energy Saving Potential in Indian Households from Improved Appliance Efficiency. (Prayas Energy Group 2010)
- A World Bank Study – Strategies for Low Carbon Growth in India: Industry and Non-Residential Sectors, prepared by LBNL. (Sathaye, S. de la Rue du Can et al. 2010).
- "Potential benefits from improved energy efficiency of key electrical products: The case of India." Lawrence Berkeley National Laboratory, Energy Policy 2008 (precursor to this report). (McNeil, Iyer et al. 2008)
- A recent study by LBNL assessing efficiency improvement potential in the residential sector Coping with Residential Electricity Demand in India's Future - How Much Can Efficiency Acheive? – Lawrence Berkeley National Laboratory 2007. (Letschert and McNeil 2007)

This study seeks to provide policymakers and other stakeholders with actionable information towards a road map for reducing energy consumption in the most cost-effective way. A major difference between the current study and some others is that we focus on individual equipment types that might be the subject of policies - such as labels, energy performance standards, and incentives - to affect market transformation in the short term, and on high-efficiency technology options that are available today.

The approach of the study is to assess the impact of short-term actions on long-term impacts. "Short term" market transformation is assumed to occur by 2015, while "long-term" energy demand reduction impacts are assessed in 2030. In the intervening years, most but not all of the equipment studied will turn over completely The 15-year time frame is significant for many products however, indicating that delay of implementation postpones impacts such as net economic savings and mitigation of emissions of carbon dioxide. Such delays would result in putting in place energy-wasting technologies, postponing improvement until the end of their service life, or potentially resulting in expensive investment either in additional energy supplies or in early replacement to achieve future energy or emissions reduction targets.

1.1. Policies and Programs to Encourage Efficiency

Over the last 10 years, India has been increasingly paying attention to developing an efficient climate change policy. This chapter first describes the framework policies that have been implemented in India to promote energy efficiency actions, notably, the Energy conservation Act of 2001 and the National Mission on Enhanced Energy Efficiency (NMEEE) in India. Following, a description of government actions that concerns the building sector is given. This encompasses appliance and equipment standards and labeling and building codes. Finally, actions that are pursued either by private investors or utilities to increase the penetration of more efficient appliances are explained.

Energy Conservation Act 2001

The Energy Conservation (EC) Act, signed in 2001, provides the legal and institutional framework for the government of India to promote energy efficiency across all sectors of the economy. A coordinating body called the Bureau of Energy Efficiency (BEE) was created to implement the EC Act. Furthermore, the Energy Conservation Act was recently amended (2010) to empower BEE to accredit energy auditors and to hire its own staff, and to empower the Central Government to issue energy savings certificate. The need to improve energy efficiency was further emphasized in the National Action Plan on Climate Change (NAPCC), adopted in 2008.

National Mission for Enhanced Energy Efficiency (NMEEE)

Recognizing the importance of addressing issues related to climate change, as well as considering economic and social developmental as priorities, India outlined domestic actions towards climate change mitigation in its National Action Plan for Climate Change in 2008. The National Action Plan contain 8 National Mission that represent multi-pronged, long term and integrate strategies for achieving key goals in the context of climate change. These Missions are:

- National Solar Mission,
- National Mission on Enhanced Energy Efficiency,

- National Mission on Sustainable Habitat,
- National Water Mission,
- National Mission for Sustaining the Himalayan Eco-system,
- National Mission for a Green India,
- National Mission for Sustainable Agriculture and
- National Mission on Strategic Knowledge for Climate Change.

Each National Missions is institutionalized by a respective Ministry. The National Mission for Enhanced Energy Efficiency (NMEEE) operates under BEE. The Prime Minister's Council on Climate Change approved draft principles of the NMEEE on August 2009 and the Union Cabinet approved its implementation framework on 24th June 2010 with dedicated funds in tune with Rs. 235.35 crores (53 million US$)[1].

The NMEEE contains four initiatives:

The most advanced is Perform Achieve and Trade Scheme (PAT). PAT intends to create a market where large industries can value and exchange energy savings achieved.

The second initiative is Market Transformation for Energy Efficiency (MTEE), which includes accelerating the shift to energy efficient appliances. This initiative includes a list of recommended actions: designing a national CDM roadmap, developing CDM DSM programs (such as the Bachat Lamp Yojana described in the next sections), expending the scope of standards and labeling, organizing public procurement, developing technology programs, reinforcing Energy Conservation Building Codes (ECBC), promoting ESCOs, and encouraging capacity building and information[2]. MTEE also includes an initiative called SEEP (Super Efficient Equipment Program) which envisages developing equipment 50% more efficient than five star appliances. Ceiling fans have been chosen as the first product for this program, with a power consumption of 20-30 watts instead of 50 watts from the five star rating fans. It is planned that an incentive would be payable for every SEEP fan sold by the manufacturers. This could then be further extended to other products like Television in collaboration with other countries.

NMEEE's third initiative is the Energy Efficiency Financing Platform (EEFP). EEFP main goal is to facilitate energy efficiency project financing by engaging bank and investor to fund ESCOs.

Finally, the Framework for Energy Efficient Economic Development (FEEED) has a goal to develop two funds that will be used to guarantee ESCO repayment - the Partial Risk Guarantee Fund (PRGF) and Venture Capital Fund for Energy Efficiency (VCFEE).

1.2. Regulatory Actions

Standards and Labeling Programs

Standards and labeling (S&L) programs have been identified as one of the key activities for energy efficiency improvement. In 2006, BEE launched the National Energy Labeling Programme. The program was initially launched on a voluntary basis for two appliances, frost-free refrigerators and tubular fluorescent lamps.

In 2010, the labeling became mandatory for these two appliances in addition to air conditioners and distribution transformers while being voluntary for direct cooling refrigerators, induction motors, pump sets, ceiling fans, LPG stoves, storage water heaters (electric geysers), color televisions and washing machines.

The label is a comparative label based on 5-star rating system with the annual or daily energy consumption given to allow comparison between models. More recently BEE began actions to include laptops and printers in its star labeling program. Standards for minimum energy consumption have recently (2010) been adopted for 3 products: room air conditioners, domestic refrigerator - frost free & direct cool refrigerator and distribution transformers[3].

Building Codes

In 2007, BEE issued the Energy Conservation Building Code (ECBC) which provides minimum energy performance standards for design and construction of commercial buildings with a connected load of 500 kW and above. ECBC takes into account the five major climatic regions of India and is currently a voluntary program. However, a number of states have recently announced that they will adopt it as a mandatory requirement.

In order to accelerate the energy efficiency activities in the commercial buildings, BEE also developed a Star Rating Program for office buildings. The energy performance of a building is measured in terms of annual electricity usage per unit of built up area (in kWh/m2/year). Office buildings are rated on a 1-5 star scale taking into account building type, climate and percentage of building area that is air-conditioned, with a 5-star rating being the most energy-efficient (Kumar, Kapoor et al. 2010).

1.3. Voluntary Programs

Bachat Lamp Yojana

In February 2009, BEE launched the Bachat Lamp Yojana (BLY) program. The program aims at providing CFLs to grid-connected residential households in exchange of an incandescent lamp and for the price of an incandescent bulb (i.e. Rs 15). The scheme works on a voluntary basis and is a public-private partnership between the Government of India, private sector CFL suppliers and State level Electricity Distribution Companies (DISCOMs) which leverages the Clean Development Mechanism (CDM) of the Kyoto Protocol to recover the cost differential between the market price of the CFLs and the price at which they are sold to households.

Instead of registering separate CDM projects, a programmatic project has been registered by BEE in 2010 to UNFCCC on behalf of the country. BEE acts as a program coordinator by developing CDM methodology, monitoring and facilitating verification of certified emission reductions. CFL suppliers secure financing of initial investment for the cost differential and provide quality CFL. DISCOMs then distribute a maximum of two CFL per household in exchange of incandescent bulbs and insure installation[4].

As of Feb 2011, the program has been successful in the state of Kerala where 13 million CFLs have already been distributed, leading to a reduction of 230 MW in peak power consumption. Similar projects are also under implementation in other States including Punjab and Karnataka (Mathur 2011).

Utility Financial Incentives

Electricity efficiency programs in India are driven by the need to solve the problem of electricity shortage. Power shortages represent on average 9% of electricity demand and 14% of peak load (G. Pandian 2008). An additional 100 GW capacity is needed by 2012. In this context, energy efficiency options are among the least-cost options to mitigate the gap between demand and supply.

Utility programs in India are voluntary. There is no national prerequisite or state regulation requiring a utility to implement energy efficiency programs. Power sector reform in India started in the early 90's, and the sector is still experiencing reforms. The restructuring started with the introduction of private investment in the supply side. Electricity management is organized at the state level. Most of the states have constituted a State Electricity Regulatory Commission and in 2007, about 14 had unbundled their state electricity company (The Energy Research Institute (TERI) 2008),(Singh 2006) Along with this current trend in reform are additional debates to further broaden and deepen the process to institutionalize DSM in energy planning.

Several utilities have implemented some pilot DSM programs. Ahmadabad Electric Company (AEC) has been a pioneer in this domain. It is the largest private electric utility in India by serving 820,000 customers (Weisbrod, Tribble et al. 1998) Today the utility has several DSM programs among which is included the leasing of energy efficiency equipment to consumers with the help of an ESCO (Prayas Energy Group 2010). Consumers pay for the capital cost over a long period through the achieved savings.

The Maharashtra state electricity distribution company recently implemented a DSM pilot program to replace incandescent bulbs with compact fluorescent lamps. Consumers repay the initial cost through the savings achieved by the use of the CFL over a nine month period in utility bills (Singh, Sant et al. 2007).

Based on an analysis of the economic potential for DSM programs, regulators in the state of Maharashtra announced plans to pursue DSM programs (Phadke, Sathaye et al. 2005). The Maharashtra Electricity Regulatory Commission (MERC) issued an order in April 2008 for utility companies to pursue DSM programs. Current proposals focus on commercial building retrofits, switching from inefficient T-12 to T-5 fluorescent lamps, new commercial buildings, municipal water pumping, etc. Similar programs are being developed in other states in the country such as Delhi.

2. Energy Demand Scenarios

As the study title suggests, the high efficiency or *Business Case* scenario is constructed around a model of cost-effective efficiency improvement. The point of the study is to demonstrate that a significant reduction in energy consumption and emissions is achievable at a net negative cost, that is, as a profitable investment for society. There are a variety of ways of assessing costs and benefits to society. We chose to focus on the end user's perspective: costs in terms of additional retail equipment prices (capital investments); savings from reduced energy bills (operating costs). Only direct energy savings are included, without valuing non-energy benefits that may also accrue (comfort, productivity, health). Finally, the cost-benefit analysis is made without the elevated effective energy prices that could be implied by carbon taxes, carbon trading schemes or other policies. Savings relative to the base case as calculated in this way is often referred to as "economic savings potential".

A national-level high-efficiency scenario is constructed by assuming that market transformation to high-efficiency technologies will occur by 2015, which is judged to constitute the "short term" by the study, because it considers that five years is sufficient time to achieve market transformation through aggressive policies and stakeholder actions. The study does not model specific actions, which could include mandatory standards, voluntary labeling programs, voluntary agreements by manufacturers, utility demand-side-management programs and others[5].

The target efficiency level chosen is that which *maximizes efficiency while providing a net benefit to consumers*. This is to be contrasted with scenarios which maximize consumer payoff but not necessarily efficiency improvement, or those that include the best available technology ("max tech") without consideration of cost-effectiveness. Consumer cost-benefit analysis is evaluated in terms of cost of conserved energy. Cost of conserved energy (CCE) is the amortized incremental cost of equipment divided by annual energy savings. In other words, it's the additional annual capital investment needed to purchase high-efficiency equipment instead of bascline equipment, divided by the energy savings provided by the investment. This quantity, which has units of USD per unit energy, can be compared to prevailing energy prices to assess consumer cost-effectiveness. Technologies with a CCE less than forecast energy prices in 2015 are deemed cost-effective.

A few comments about whether this definition is optimistic or pessimistic are warranted. On one hand, high efficiency technologies are compared to the current baseline technology, even though there may already be a market for higher efficiency equipment, and the average efficiency of the market is constantly improving. This tends to underestimate the baseline forecast and overestimate savings. On the other hand, it likely underestimates the efficiency that will be achievable in a cost-effective way, first of all because technology costs are generally decreasing (according to technological learning rates) and the emergence of new technologies that may not be available for analysis. Therefore, there are two compensating effects not taken into account in the analysis. The results should therefore be taken as representative of the scale of potential improvement, not as a reliable prediction. The methodology is chosen to maximize concreteness and defensibility by relying on technologies that can be justified by actual cost data.

2.1. Literature Review

Some recent examples of studies that have identified potential energy savings from energy efficiency improvements include:

China

- China's appliance standards are estimated to have saved 1.08 EJ during 2006-2008, with refrigerators, air conditioners and televisions contributing the bulk of the savings. (Price, Levine et al. 2011)
- (Fridley 2008) estimates potential savings of 1.2 TWh in 2012 and 16 TWh by 2020 for energy labels on refrigerators in China.
- (Cheung and Kamg 2008) describe the growth of China's energy efficiency industry, projecting spending of USD 300 billion over five years.
- (Aden, Qin et al. 2010) uses lifecycle assessment to show that for buildings in the Beijing area, 80% of energy use and related emissions is due to operations, and about 20% due to materials.
- (Zhou 2010) provides an overview of China's policies on energy efficiency.

India

- (Delio, Lall et al. 2009) estimates potential savings from energy efficiency across all sectors in India to be 183 TWh in five years.
- (de la Rue du Can 2009) provides both retrospective and prospective views of energy use in the residential and transport sectors of India.

United States

- The National Research Council report, *America's Energy Future,* in 2009 estimated potential cost-effective energy savings in the U.S. of about 20% in 2020 and about 30% in 2030, with the greatest potential in the buildings sector (National Research Council, *Limiting the Magnitude of Future Climate Change*, 2010).
- The American Physical Society report, *Energy Future: Think Efficiency* (2008) estimated 572 TWh of electricity savings in the residential sector in 2030, and about 30% savings for the building sector as a whole, all below the retail price of electricity energy.
- The U.S. Department of Energy's Appliance Standards Programs has conducted extensive studies for regulated product types

(http://www1.eere.energy identifying economically justified and technologically feasible energy efficiency improvements.
- The Energy Information Administration annually publishes additional efficiency scenarios, e.g., high technology cases, in conjunction with the *Annual Energy Outlook* (http://www.eia.doe.gov/oiaf/aeo/).

2.2. Construction of the Energy Demand Scenarios

Any study that aims to project energy efficiency improvements from specific technologies must make the link between unit-level improvements and national impacts. The current study achieves this using LBNL's Bottom-Up Energy Analysis System (BUENAS). As the name suggests, BUENAS is a bottom-up technology-oriented model, rather than a top-down macroeconomic model[6]. BUENAS combines unit-level efficiency scenarios with a forecast of stock size and turnover to calculate national energy savings impacts through 2030. Unit level energy demand by baseline and "target" technologies are collected in a database that the model takes as inputs, and which define the base case and high efficiency scenarios. Growth of the stock (number of units operating) by 2030 is a function of economic and population growth.

BUENAS uses minimum efficiency performance standards (MEPS) as a default policy, that is, it models a discrete change in the efficiency of equipment after a specific year. For the current study, we chose an implementation year of 2015, assuming that several years lead time are necessary between identification of efficiency targets, and making them mandatory.

Originally constructed as a global model, BUENAS covers a wide range of energy-consuming products, including most appliance groups generally covered by Energy Efficiency Standards and Labeling (EES&L) programs around the world. The global model covered the following appliance groups:

- *Residential Sector:* Lighting, Refrigerators, Air Conditioners, Fans, Washing Machines, Standby Power, Televisions, Electric Ovens, Space Heating and Water Heating.
- *Commercial Building Sector:* Lighting, Air Conditioning, Refrigeration, Ventilation, Office Products, Space Heating and Water Heating.
- *Industrial Sector:* Electric Motors.

For the purposes of the *India Business Case for Energy Efficiency*, many of the end uses needed for the analysis were present in BUENAS. However, some modifications were made. First, the *Business Case* model is dependent on an evaluation of cost-effectiveness. Therefore, appliance groups for which data were insufficient to permit this calculation were not included. On the other hand, some equipment types for which data were available were not included in the original model, such as distribution transformer. In that case, this end use was added. While residential fans are a very important end use in India, cost data for this appliance were not available at the time of the analysis, so they are omitted. Finally, televisions are not covered here. While efficiency improvements are certainly possible in television displays, this is a dynamic and rapidly evolving technology. Recent market trends include a massive shift to flat panel technology, with dramatic increases in screen size, along with market-driven efficiency improvements. Because of the dynamism in television technology, efficiency baselines and technology trends cannot be adequately forecast in order to provide definitive cost-effective improvement potential.

The BUENAS model uses the Long Range Energy Alternatives Planning (LEAP) platform[7] to forecast energy consumption by end use from 2005 (base year) to 2030. The strategy of the model is to first forecast end use *activity*, which is driven by increased ownership of household appliances and growth in the industrial sector. The total stock of appliances can be modeled either according to an econometric diffusion model or according to unit sales forecasts, if available. Electricity consumption or *intensity* of the appliance stock is then calculated according to estimates of the baseline intensity of the prevailing technology in the local market. Finally, the total final energy consumption of the stock is calculated by modeling the flow of products into the stock and the marginal intensity of purchased units, either as additions or as replacements of old units. The high efficiency or "policy" scenario is created by the assumption of increased unit efficiency relative to the baseline starting in a certain year. For example, if the average baseline unit energy consumption (UEC) of new refrigerators is 450 kWh/year, but a MEPS taking effect in 2012 requires a maximum UEC of 350 kWh/year, the stock energy in the policy scenario will gradually become lower than that of the base case scenario due to increasing penetration of high-efficiency units under the standard. By 2030, the entire stock will generally be impacted by the standard[8]. Figure 1 shows the analytical structure of BUENAS.

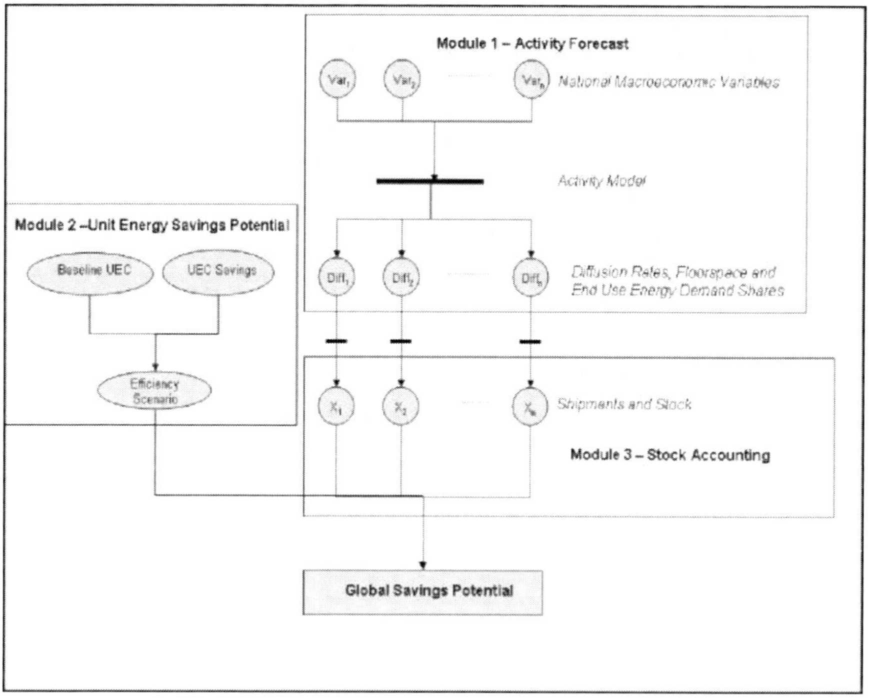

Figure 1. Structure of BUENAS.

The main outputs of BUENAS are base case energy consumption forecasts to 2030 by end use and energy, energy saving impacts of the modeled policy, and carbon dioxide emissions mitigation impacts. For this study, financial impacts were added to the model in a spreadsheet calculation.

For the residential sector, *activity* as modeled in Module 1 of the model is given by the stock of equipment, that is, the number of appliances installed and operating in Chinese households in a given year.

Once the number of residential products in each appliance group in each year is established, this number is multiplied by the *annual unit energy consumption* (UEC) to yield energy demand for the appliance group. *UEC* is the subject of Module 2 of BUENAS, and determines the efficiency scenario modeled. Determination of the baseline and efficiency scenario UEC is discussed in Section 3 below.

Finally, Module 3 tracks the introduction of each year's cohort of appliances into the stock, taking account of growth in the market, equipment retirements, and replacements. Retirement and survival functions are derived from average lifetimes and assumed to have a distribution around the mean value. This shape of the retirement function is assumed to be that of a normal distribution centered around the mean lifetime by default, but takes the form of a more complicated function (Weibull distribution) if such a distribution is available. The survival function is given by:

$$Survival(age) = 1 - \int Retirement(age)$$

Using the retirement distribution, the model calculates the weighted average efficiency of the stock in each year. In the case of the high efficiency scenario, only a small fraction of the stock operates at high efficiency in the years immediately following the policy start date, but this fraction grows over time. The percentage of stock operating in 2030 that was installed after the policy start date is dependent on the assumed average lifetime of the product class.

Compared to the residential sector, energy demand in the commercial building sector is driven by a much wider variety of equipment types and follows distinct usage patterns depending on the type of building. For this reason, BUENAS models commercial buildings in an aggregate fashion, rather than at the level of individual appliances. In Module 2, the commercial sector model uses aggregate energy intensity numbers for major appliance categories, such as lighting, space heating and air conditioning and refrigeration. In order to model energy demand and savings from efficiency improvement, we estimate the fraction of energy covered by individual technologies for which data are available. Energy and demand are thereby calculated from base year values of energy intensity according to a scaling factor.

3. Efficiency Improvement Potential – Cost-Benefit Analysis

Cost-effectiveness is defined in terms of cost of conserved energy, that is, how much the end user must pay in terms of annualized incremental equipment investment for each unit of energy saved by higher efficiency equipment. The formula for cost of conserved energy is

$$CCE = \frac{I \times q}{S} \qquad (1)$$

In this equation, I is the total additional investment needed to purchase high efficiency equipment rather than the baseline technology, and S is the resulting annual energy savings. The capital recovery factor q is given in turn by:

$$q = \frac{d}{(1-(1+d)^{-L})} \quad (2)$$

In this equation, d is the end user discount rate and L is the average lifetime of the equipment, in years. Defined in this way, I times q is an annual payment for an amortized capital investment. Cost of conserved energy is a convenient metric for comparison of cost-effectiveness of measures[9].

3.1. Equipment Data

Since we know of no systematic database of efficiency and cost of energy-using equipment in India, the evaluation of cost-effectiveness of Indian efficiency technologies for this study relied on a variety of sources. Identification of efficiency for refrigerators and air conditioners was facilitated by the mandatory Energy Label program and studies related to it (Tathagat and Anand 2011). In the case of lighting and other commercial equipment, we drew on data collected for a previous study performed at LBNL for the World Bank (Sathaye, S. de la Rue du Can et al. 2010). Likewise, the distribution transformer analysis utilizes data from an earlier study (McNeil, Iyer et al. 2008). Finally, we find industrial motor efficiency improvement to be cost effective base on pricing data from the United States. Assumptions of baseline energy consumption, high efficiency levels and price data sources are shown in Table 1.

Incandescent Lamps

Replacement of incandescent lamps with compact fluorescent lamps (CFLs) or other technology such as LEDs is generally at the top of the list of attractive efficiency measures because of the large fractional savings (up to 75%) and the high degree of cost-effectiveness. Bans on incandescent lamps are also among the most popular efficiency policies globally.

Indeed, due to Bachat Lamp Yojana and other government programs, we expect increased penetration and eventual phase out of incandescent lamps in favor of CFLs in the base case scenario. In this case, the high-efficiency scenario is characterized by an acceleration of existing policies leading to a complete phase out of incandescent after 2015. In order to model energy savings, we assume that the typical incandescent lamp in India is 60W and is operated for 4 hours per day on average in the residential sector and 8 hours per day in the commercial sector, for an annual energy consumption of 87.6 kWh and 175 kWh respectively. An equivalent CFLs is assumed to use only 15W, or 21.9 kWh and 43.8 kWh per year respectively. We assume that a CFL lasts for 5 years, compared to only 1 year for incandescent lamp.

Fluorescent Lamps

Fluorescent tube lights account for approximately 43% of the lighting fixtures in residences (Prayas Energy Group 2010) and roughly 70% of lighting in the commercial sector (Sathaye, S. de la Rue du Can et al. 2010) in India. The baseline fluorescent ballast is taken to be a 40 watt T12 lamp coupled with a magnetic ballast, and is estimated at 46W total. For a high-efficiency option, we consider a high performance T8 with an electronic ballast, estimated at 41W. We assume fluorescent lamps operate for 4 hours per day in the residential sector and 8 hours per day in the commercial sector. With these estimates and assumptions, baseline fluorescent lamp-ballast combinations consume 67.2 and 134.3 kWh per year in the residential and commercial sectors respectively, while high-efficiency combinations consume 59.8 and 119.5 kWh respectively. The lifetimes of fluorescent ballasts are assumed to be 15 years.

Refrigerators

Estimates for refrigerator improvement potential are based on the current BEE labeling scheme. Current market shares indicate that the bottom of the market is roughly at the 4-star level for direct cool refrigerators and at 3 stars for frost-free refrigerator-freezers (Tathagat and Anand 2011). For high-efficiency, we consider a target of 5-stars[10]. As determined by the labeling definitions, we estimate baseline energy consumption to be 337 kWh for direct cool refrigerators and 675 for frost-free. The 5 star target corresponds to 269 and 432 kWh respectively. In modeling national energy savings from refrigerators, we include a shift from direct cool refrigerators to frost-free units, from 15% frost free in 2005 to 75% in 2030. We assume that a refrigerator lasts 15 years (McNeil, Iyer et al. 2008).

Room Air Conditioners

Room air conditioners are also covered by the 5-star BEE labeling scheme. The Indian room air conditioner market includes both window and split type units, with a somewhat larger share for split units. Roughly speaking, the room air conditioner market was found to be at the 2-star level for window units and 3 stars for split units, with about 16% of the overall market still at the 1- star level. Market weighted efficiency for both types of units is 2.61[11]. Air conditioner annual energy consumption is calculated assuming a 6 month cooling season and 8 hours per day use (McNeil and Iyer 2009). The typical unit is assumed to be a 1.5 ton unit operating at 75% capacity, which yields 2160 kWh per year for residential consumers.

As in the case of refrigerators, the analysis is limited to considering the 5-star case due to data considerations. This corresponds to an EER of 3.1 W/W. Scaling the baseline efficiency by energy efficiency yields a target energy consumption of 1812 kWh per year. Room air conditioners also account for roughly 30% of mechanical cooling in commercial buildings. These are considered in the commercial air conditioning section below. The average lifetime of Room Air Conditioners is assumed to be 15 years (Letschert and McNeil 2007).

Standby Power

Standby power consumption is a feature of a wide range of products, including major appliances, consumer electronics and home entertainment equipment. This mode of power consumption is increasingly shown to be a major source of energy demand, and has become a prominent candidate for efficiency improvement (IEA, 2001). Reduction of standby power is typically very inexpensive to achieve through redesign of electronic components. Standby power savings potential is modeled after the *Preparatory Study* for recent regulations in the European Union (EC 2007). According to that study, the average product consuming standby used 17.2 kWh in this mode at the time of the study. This consumption was reduced by the EU's Tier 1 standard to 7.1 kWh in 2010 and is set for another reduction, to 3.6 kWh in 2013. In the Indian Business Case scenario, we assume that the EU Tier2 level will be reached by 2015. We assume that the average product using standby power lasts 8 years.

Commercial Lighting

Commercial lighting in India is provided primarily by linear fluorescent lamps, incandescent lamps and CFLs. Data provided for a recent LBNL study

(Sathaye, S. de la Rue du Can et al. 2010)indicates a 72% / 27% / 7% split between these three main lighting types. Data for high-intensity discharge lamps and other lighting types were not available, so these are not considered. Commercial lighting baseline and target efficiencies are assumed to be the same as for the residential sector. These entail a switch from T12 to T8 and magnetic to electronic ballasts for fluorescent tubes and switching the remaining incandescent lamps for CFLs.. The commercial sector case assumes 8 hours per day lighting usage, however, instead of 4.

Commercial Air Conditioning

Air conditioning is provided by a variety of technologies in Indian commercial buildings, including central air conditioners, heat pumps and chillers, in addition to room air conditioners, which are still the most common form of air conditioning. The evaluation of commercial air conditioning potential relies on data collected for a recent LBNL study (Sathaye, S. de la Rue du Can et al. 2010), which estimates incremental equipment costs and energy savings from high-efficiency equipment. Overall improvement potential and net costs are calculated using weighted averages over technology and building types.

Industrial Motors

The Indian motor efficiency labeling system is modeled after the original European scheme that defines the highest efficiency motors as IE3, less efficient motors above a standard baseline as IE1 and IE2. Currently, much of the Indian motors market remains below the IE1 level, and there is a small fraction of motors at the IE2 level[12]. However, we expect the market to reach roughly the IE1 level, either through market/voluntary forces or mandatory regulations (Garg 2009). We consider IE1 as a target efficiency level, and assume a constant ratio of IE2 to IE1.

Annual energy consumption for motors is based on estimates from the Preparatory study for recent Ecodesign standards in the E.U. (de Ameida, Ferreira et al. 2008). This study takes into account the variation in motor usage with application and size. We consider three categories of motors: 0.75-7.5 kW, 7.5-75 kW and over 75 kWh. The market average of these categories used in the EU study is 1.1 kW, 11 kW and 110 kW respectively. Average UEC are found to be 1485 kWh, 149800 kWh and 396000 kWh. These estimates show a higher intensity for larger motors in terms of load and hours of use. The distribution of motor capacities in is assumed to be in a 100:10:1 ratio for the three categories. The lifetime for all motors is assumed to be 10 years.

Distribution Transformers

We analyze distribution transformers according to the star rating defined by the Bureau of Energy Efficiency (BEE). Within this scheme, cost-effective efficiency improvement for distribution transformers was considered in a previous study (McNeil, Iyer et al. 2008). Standards proposed at the time of that study are currently still in effect[13]. While a requirement of Level 1 is nominally in effect, there is a very large amount of 'leakage' in the program and in fact only a small fraction of transformers in India are labeled[14]. Therefore, we consider the lowest BEE rating (Star 1) as the baseline, and consider the cost effectiveness of a market shift to the highest rating (Star 5). Distribution transformer lifetime is assumed to be 22 years.

Average lifetime, baseline and target unit energy consumption are summarized in Table 1. Sources for these data are summarized in the Appendix.

Table 1. Lifetime and Energy Parameters for Indian End Uses

Equipment	Lifetime	Baseline UEC	Target UEC
Electric Equipment	years	kWh	
Incandescent Lamps (Residential)	5*	88	22
Incandescent Lamps (Commercial)	5*	175	44
Fluorescent Lamps (Residential)	15	67	60
Fluorescent Lamps (Commercial)	15	134	120
Refrigerators (Direct Cool)	15	277	221
Refrigerators (Frost Free)	15	486	311
Air Conditioners (Residential)	15	2160	1812
Standby Power	8	17	4
Commercial Cooling	15	-	1012**
Motors			
10HP (90%)	10	1485	1395
50 HP (9%)	10	19800	19365
100 HP (.9%)	10	396000	392234
Distribution Transformers			
25 kVA (6%)	22	1036	441
60 kVA (25%)	22	1834	797
100 kVA (44%)	22	2619	1068
160 kVA (3%)	22	3757	1653
200 kVA (14%)	22	4989	1880

* Lifetime of CFL replacement.

** Only energy savings available.

3.2. Cost of Conserved Energy Calculation

Prices used for the calculation of cost of conserved energy are given in Table 2. Sources for these are provided in the Appendix. These parameters are used in the calculation of cost of conserved energy according to Equation 1 by comparing each design option to the baseline, according to:

$$I = Price_{DesignOption} - Price_{Baseline}$$

and

$$S = UEC_{Baseline} - UEC_{DesignOption}$$

The parameters used in calculation of q in Equation 2 are as follows:

Product Lifetime (*L*) – Average number of years that a product is used before failure and retirement. Lifetimes vary by product class and are estimated from manufacturer reports, or from survey data.

Discount Rates – In order to evaluate cost-effectiveness to consumers, the analysis takes into account the real cost of financing for Indian consumers. We assumed a *real discount rate* of 10% for all consumers. This rate is intended to represent actual interest rates on the financing of equipment, and does not include consumers' high sensitivity to first costs, such as an *implicit discount rate*. This discount rate is high compared to rates used in appliance efficiency studies in the United States which use rates of about 5% for residential and 6% for commercial consumers (McNeil, Bojda et al. 2011).

Energy Prices – Current consumer electricity and natural gas prices in India generally do not fully reflect the cost of production and include cross-subsidization between tariff groups. However, in recent years, Indian electricity tariffs have been increasing as subsidies are lowered. We estimate electricity prices based on state-wise/utility wise average rates of electricity for domestic and industrial consumers[15]. In order to make an estimate of the national level, state-level tariffs are weighted by state populations[16]. Using this method, average residential rates from 2009-2010 were found to be 3.49 INR/kWh, or $0.0714 USD/kWh. Industrial rates were 4.47 INR/kWh or $0.0915/kWh. Commercial rates were not available – industrial rates were uses as a proxy for those. We assume that electricity rates remain constant at these levels, an assumption that is likely conservative.

Using these parameters, we calculate cost of conserved energy for each design option for each product class. The results of this calculation, shown in Table 2, are the basis of construction of the efficiency scenario.

As stated above, the target efficiency level chosen is that which *maximizes efficiency while providing a net benefit to consumers*. Following this definition, we identify the target UEC for each product class as the lowest UEC for which cost of conserved energy is below the utility price.

To illustrate the construction of the efficiency scenario, we consider the example of frost-free refrigerators. Table 2 shows *UEC*, *Price* for baseline (3 Star) and high-efficiency (5 Star) refrigerators. According to a retrospective study of program impacts in 2010 (Tathagat and Anand 2011), a 5 Star frost-free refrigerator costs 2000 INR or $41 USD more than a similar 3 Star unit. Annual Unit Energy Consumption (UEC) savings of this substitution is 175 kWh per year. With a discount rate of 10% and a lifetime of 15 years, equation 2 gives a q value of 0.131. Cost of conserved energy is therefore given by

$$\frac{\$41 \times 0.131}{243\ kWh} = 0.022\ \$/kWh$$

Calculation parameters and results for cost-of conserved energy for all products studied are shown in Table 2. References and assumptions for energy, efficiency and price parameters for appliances are presented in the Appendix.

Table 2. Cost of Conserved Energy Calculation

Equipment	Lifetime	q	Baseline UEC	Target UEC	Baseline Price	Target Price	Target CCE
Electric Equipment	years		kWh		$	$	$/kWh
Incandescent Lamps (Residential)	5*	0.264	87.6	21.9	0.29	4.11	0.011
Incandescent Lamps (Commercial)	5*	0.264	175	43.8	0.29	4.11	0.005
Fluorescent Lamps (Residential)	15	0.131	67	59.8	-	3.46**	0.062
Fluorescent Lamps (Commercial)	15	0.131	134	119.5	-	3.46**	0.031
Refrigerators (Direct Cool)	15	0.131	337	269	-	20**	0.040
Refrigerators (Frost Free)	15	0.131	675	432	-	41**	0.022

Table 2. (Continued)

Equipment	Lifetime	q	Baseline UEC	Target UEC	Baseline Price	Target Price	Target CCE
Electric Equipment	years		kWh		$	$	$/kWh
Air Conditioners (Residential)	15	0.131	2160	1817	-	77**	0.035
Standby Power	8	0.187	17.2	3.6	-	2**	0.028
Commercial Cooling	15	0.131	-	1012**	-	178**	0.072
Motors							
10HP (90%)	10	0.163	1485	1395	100	125	0.046
50 HP (9%)	10	0.163	19800	19365	403	575	0.065
100 HP (.9%)	10	0.163	396000	392234	2858	3991	0.049
Distribution Transformers							
25 kVA (6%)	22	0.114	1036	441	1036	441	0.065
60 kVA (25%)	22	0.114	1834	797	1834	797	0.051
100 kVA (44%)	22	0.114	2619	1068	2619	1068	0.037
160 kVA (3%)	22	0.114	3757	1653	3757	1653	0.016
200 kVA (14%)	22	0.114	4989	1880	4989	1880	0.030

* Lifetime of CFL replacement.
** Only incremental price or energy savings available.

The cost of conserved energy for all appliance groups is compared to utility prices in Figure 2. It is notable that in general, the results are not highly sensitive to assumptions about future electricity price, since the efficiency targets studied all fall well below current electricity prices. This result is indicative of two major factors. First, the efficiency technology baseline of India is relatively low.

This generally means that there is more 'low hanging fruit' still available to be captured in the Indian context. Second, however, we considered only those technologies for which prices were available and could be evaluated in the Indian market[17].

The main inputs to the construction of the two scenarios, the *Base Case* and the *Business Case* scenario are the baseline *UEC* and the UEC established by CCE in Figure 2. We call this the *Business Case UEC*.

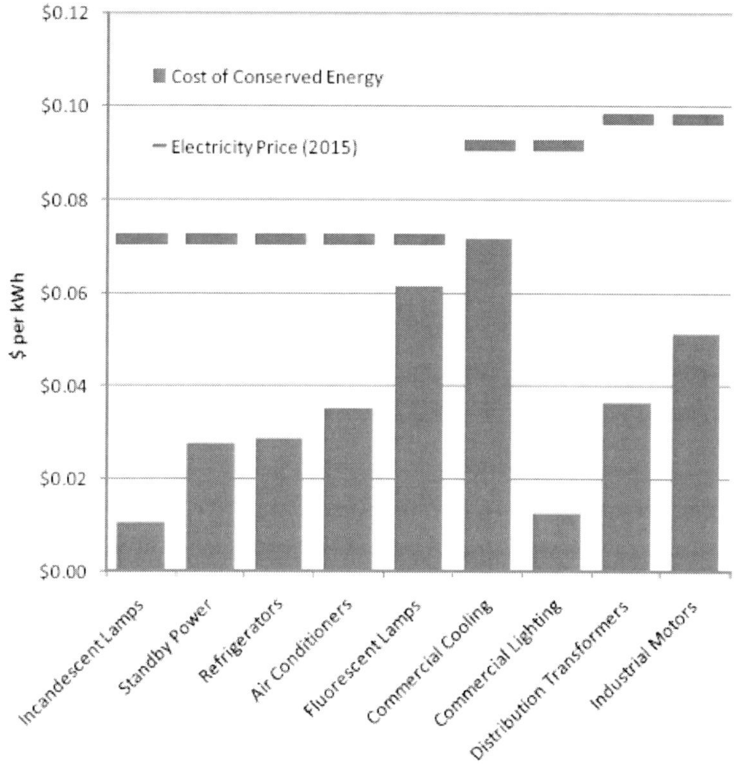

Figure 2. Cost of Conserved Energy and Energy Prices.

4. National Level Energy Savings Opportunities

Because of the modular structure of the BUENAS model (see Figure 1), once the inputs are established it is a relatively straightforward process to construct the two energy demand scenarios and compare them to calculate savings potential. The full details of the calculation of energy demand are provided in (McNeil, Letschert et al. 2008) and are omitted here.

4.1. Energy Savings and Emissions Reductions

Site energy savings is the basis for all national impacts calculations. Site energy demand refers to electricity and natural gas consumed in a home or business, and does not include fuel inputs in generation of electricity, or losses in transmission or distribution. Site energy is the energy affected most immediately by efficiency improvement. It is also the energy consumption that appears on consumer utility bills, and forms the basis for the cost-benefit analysis detailed above.

Site energy consumption is calculated by BUENAS for both the *Base Case* and *Business Case* scenarios. Energy *activity* is the same in both cases[18], so the difference between them is driven by the trend in *marginal intensity*, that is, the UEC of products sold in each year. The UEC for the two scenarios are identical until the policy implementation date of 2015[19]. After that date, the efficiency target in the *Business Case* is the high efficiency level determined by cost-benefit analysis, while it remains at the baseline efficiency level in the *Base Case*. The difference in UEC in the two scenarios applies only to new products – in this way, the policy modeled has the structure of a minimum efficiency performance standard, and does not imply retrofits of existing equipment. By 2016 overall energy demand of stock in the *Business Case* is only slightly lower than the *Base Case*, because only one year's sales are affected by the policy. Moving through the forecast, LEAP tracks the gradual flow of high efficiency products into the stock and the retirement of less efficient ones, so that the average stock UEC gets closer to the target level. Depending on the lifetime of the product, the entire stock may not be converted by 2030, since some low-efficiency products installed before 2015 will survive. Figure 3 shows the evolution of site energy savings by appliance group. From 2015 onward, energy savings grows for all products as high efficiency products begin to penetrate the stock in the *Business Case*.

Site energy savings results are summarized in Table 3. Savings for all appliance groups totals 139 TWh in the year 2030. Cumulative savings through 2030 total 1,251 TWh.

Emissions reductions are calculated directly from energy savings according to a carbon factor. The carbon factor for electricity includes fuel inputs to generation, and accounts for transmission and distributions losses. The carbon factor taken in the base year 2005 is estimated in (Price, S. de la Rue du Can et al. 2006). The forecast of carbon factor is derived using the base year data, and scaling by the growth rate from IEA's World Energy Outlook (WEO) 2006 (International Energy Agency 2006). Using this method, the carbon factor in India is found to be 1.12 kg CO_2/kWh in 2015, decreasing to 10.06 kg CO_2/kWh in 2030. Carbon factors for natural gas and fuel oil are assumed to remain constant at 0.202 and 0.264 kg/CO_2, respectively. Emissions reductions from energy savings determined by multiplying energy savings by carbon factors are shown in Table 3. Total mitigation in the *Business Case* is found to be 146 mt CO_2 in 2030 and 1350 mt CO_2 over the entire forecast. Figure 4 shows the contribution to cumulative CO_2 mitigation from all appliance groups.

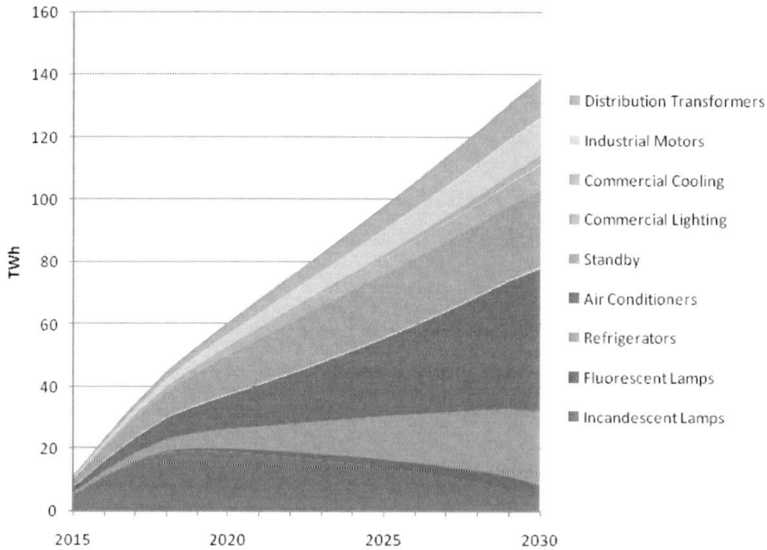

Figure 3. Site Energy Savings by Appliance Group– 2015-2030.

Table 3. Energy Savings and Pollutant Mitigation by Appliance Grou

End Use	Final Energy Savings			Emissions Mitigation		
	In 2020	In 2030	Through 2030	In 2020	In 2030	Through 2030
	TWh			mt CO_2		mt CO_2
Air Conditioners	11.0	46.0	315	12.1	48.6	339
Standby	12.6	25.0	249	13.9	26.4	269
Incandescent Lamps	18.5	3.6	207	20.4	3.8	226
Refrigerators	6.2	23.7	172	6.8	25.1	185
Industrial Motors	3.8	12.0	95	4.2	12.7	103
Distribution Transformers	3.7	12.6	95	4.0	13.3	103
Commercial Lighting	2.2	8.0	59	2.4	8.4	63
Fluorescent Lamps	1.5	4.9	38	1.7	5.1	41
Commercial Cooling	0.8	3.2	22	0.8	3.3	24
Total	60	139	1251	66	147	1351

Several conclusions can be drawn from Table 3 and Figure 4. First, emission reduction potential is well distributed among end uses and sectors. The largest potential exists for residential air conditioning, incandescent lamps and refrigerators and standby power, each of which could provide over 200 mt CO_2 over the forecast period. The large savings from room air conditioners is driven by a combination of both the high cooling load in India and the related rapid growth of this end use in Indian homes. Savings from incandescent lamp phase out is high cumulatively, but low in terms of savings in 2030. This is because incandescent are assumed to be phased out by that year even in the Base Case – the Business Case for lamps is an acceleration of that phase out. Finally, significant savings potential is also shown for refrigerators, industrial motors and distribution transformers.

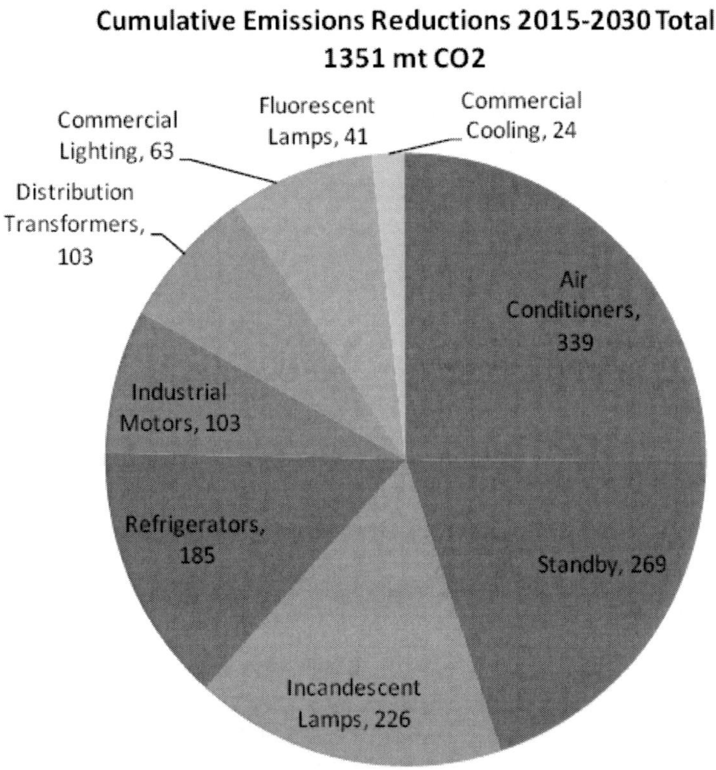

Figure 4. Cumulative CO2 Emissions Reductions 2010-2030.

4.2. Consumer Financial Impacts

By construction, the *Business Case* implements energy efficiency in a way that is cost-effective to consumers. Because this study insisted on quantifying investments needed to improve efficiency relative to the base case technology, the necessary information to evaluate these investments and financial benefits of energy savings, and therefore net financial impacts to consumers, is available for all appliance groups considered.

Recalling the definition of cost of conserved energy from Equation 1:

$$CCE = \frac{I \times q}{S}$$

The denominator of this equation *I x q* is the annualized equipment investment necessary to yield an annual energy savings *S*. BUENAS calculates the total savings *ST(y)* in each year, given by:

$$S_T(y) = S \times Stock'(y)$$

In this equation, *Stock'(y)* is the affected stock, that is, the number of units operating in the stock that were installed after the policy implementation date, and are each providing a savings *S* relative to the *Base Case*. Likewise, the total annualized investment in each year *IT(y)* x *q* is given by:

$$I_T(y) \times q = I \times q \times Stock'(y)$$

Substituting Equation 1, and cancelling terms, yields:

$$I_T(y) \times q = S_T(y) \times CCE$$

In other words, total annualized investment can be calculated for each appliance group by multiplying its total energy savings by the cost of conserved energy shown in Table 5.

Financial savings from energy savings is given simply by the utility price in each year multiplied by the total energy savings *ST(y)*. Net financial impacts are then given by:

$$N(y) = S_T(y) \times (Utility\ Price - CCE)$$

Costs, Savings and Net Impacts calculated in this way are shown in Table 5. In evaluating the financial value of efficiency or other government programs, it is customary to take account of deferred benefits through a discount rate calculation. The resulting *Net Present Value* (NPV) of benefits is given by:

$$NPV = \sum_{y=2010}^{2030} \frac{N(y)}{(1 + DR)^{y-2010}}$$

In this equation, *DR* is a "societal" discount rate that parameterizes the preference for immediate returns on public investments. We consider two scenarios in which the societal discount rate is taken to be 3% or 7%. Cumulative equipment costs, energy bill savings, net savings and *NPV* are shown in Table 5.

The analysis shows that cost-effective efficiency improvement could yield very significant financial benefits to Indian consumers. Table 4 shows positive net savings for all appliance groups, which is not surprising, since the target efficiency levels were constructed to be cost-effective. The table shows that cost-effective efficiency improvements require an investment of 38 billion USD over the next 20 years, but these investments will return over more than twice as much over the same period, for a net savings of 58 billion dollars, or of order of fifty dollars per capita.

The present value of net savings is 36 billion USD assuming a discount rate of 3%, and 22 billion USD with a 7% discount rate.

Of the appliance groups studied, residential room air conditioners require the largest investment at 11 billion USD, but provide a payoff of 22.5 billion USD. Standby power generates similar savings, but with only half of the investment.

Phasing out incandescent lamps is extremely cost effective compared to other appliance groups, with a payoff of over six times as high as the required investment.

Finally, financial impacts, emissions savings and their relationship can be shown using a "conservation supply curve". This unique way of expressing the cost and benefits of carbon mitigation measures has become very widespread in the literature because of the key information it conveys. A conservation supply curve for the *Business Case* is presented in Figure 5. The x-axis shows cumulative carbon mitigation and expresses the relative importance of each appliance group. The total extent of the curve is 1351 mt CO_2, as shown in Table 3. The y-axis displays relative affordability according to cost of conserved energy. The blocks corresponding to each measure are ordered with increasing cost of conserved energy, from left to right.

Table 4. Cumulative Financial Impacts of Efficiency Improvement

Appliance Group	Cumulative Financial Impacts				
	Cost	Savings	Net Savings	NPV @ 3% DR	NPV @ 7% DR
	Billions USD				
Incandescent Lamps	2.2	14.8	12.6	8.9	5.9
Standby	6.9	17.7	10.9	8.4	4.9
Air Conditioners	11.0	22.5	11.5	7.3	4.1
Refrigerators	4.9	12.2	7.4	4.7	2.7
Distribution Transformers	3.5	9.3	5.8	2.8	1.6
Commercial Lighting	0.7	5.4	4.6	1.6	0.9
Industrial Motors	4.9	9.3	4.4	1.2	0.7
Fluorescent Lamps	2.3	2.7	0.4	0.2	0.1
Commercial Cooling	1.6	2.0	0.4	1.1	0.6
Total	38	96	58	36	22

Finally, we note that there are other benefits to the energy savings achieved in the *Business Case* besides the direct energy and financial benefits. The effect of reduction of greenhouse gas emissions and resulting avoided costs are difficult to quantify, but could be very large. One metric to consider the order of magnitude of the value of these types of impacts is the assumption of a carbon price. The assumption of a price of 25 USD per ton of carbon dioxide yields an additional 34 billion USD of savings, while a 100 USD per ton price yields 135 billion additional USD, more than tripling the total.

The negative impacts of emissions of SO_2 and NO from power plants are well-known (see, for example, EPA 2010), including acid rain, acidification of watersheds and lakes, and respiratory illness from inhaling particulates. Likewise, the reduction of mercury emissions from coal-burning power plants reduces fish contamination, which is now recognized as a major health risk. We do not try to quantify the health impacts of reduction of these emissions, only point out the obvious – that savings due to efficiency is equivalent to installation of clean electricity generation. In the *Business Case*, this reduction provides a large net financial benefit to consumers, which may not be true with alternatives.

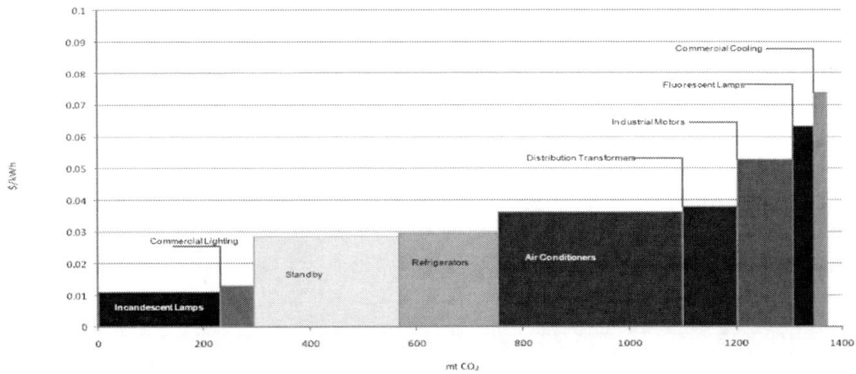

Figure 5. Conservation Supply Curve for Indian End Uses 2010-2030.

Conclusion

The *Business Case* analysis found additional potential for cost-effective efficiency improvement in India for seventeen appliance groups in the residential and commercial building sectors and industrial motors. Efficiency improvement for these technologies could deliver twice as much financial benefit to Indian households and business than the investment needed to implement them. In addition to direct financial benefits, impacts on greenhouse gas emissions and are significant. Total net impacts from additional deployment of high efficiency technology include:

Energy savings:

- 60 billion kWh per year in 2020
- 140 billion kWh per year in 2030
- A total of 1,250 billion kWh cumulatively through 2030

Cumulative greenhouse gas emissions mitigation:

- 1350 million metric tons of CO2 through 2030

Financial impacts to consumers through 2030:

- Equipment investment of 38 billion USD
- Energy bill savings of 96 billion USD
- Net savings of 58 billion USD

The "business case" analysis shows that the Indian market already has access to efficiency technologies that could provide Indian consumers with a financial benefit and make a dent in the growth of Indian emissions if widely adopted. Most of the equipment studied has been the subject of at least one efficiency standard, but opportunities for improvement are not exhausted. To some extent, therefore, the savings potential estimated by this study can be captured through expansion and aggressive pursuit of existing Indian government policies. It should also be noted that many of the technologies included in the "business case" scenario were not available ten to twenty years ago, or at least weren't be shown to be cost effective. These technologies have become available and cost-effective through research, new materials and components, improvements in production processes, or changes in design of systems. Likewise, we expect that a similar analysis performed 10 years from now will show improvements not accessible to the current study due either to lack of data or prohibitively high cost of "prototype" technologies.

Because the rigor of the methodology used to evaluate cost-effectiveness requires a significant amount of technical data, we only cover a subset of equipment types for which significant savings potential might be available. In particular, the appliance groups covered are limited to buildings applications. For this sector, however, we believe a large fraction of energy demand is accounted for. For this reason, while the overall savings potential is large, it cannot be interpreted as "comprehensive".

Finally, we believe this study to be among the few to attempt to evaluate the "economic" potential of efficiency improvement in India in a transparent way. In addition to demonstrating significant savings potential, we hope that it demonstrates a clear and consistent methodology for creation and expansion of alternative energy scenarios in India. Additional scenarios that could be explored include the potential impact of carbon taxes, cap-and-trade, R&D investments and other policy- or market-based drivers. The ability of the research community to utilize this type of analysis to inform government and private sector decision makers will depend largely on investments made in development of the type of data used here, both more widely and with greater frequency.

References

Aden, N., Y. Qin, et al. (2010). Lifecycle Assessment of Beijing-Area Building Energy Use and Emissions: Summary Findinga and Policy Applications. Berkeley LBNL.

Cheung, R. and A. Kamg (2008). China's Booming Energy Efficiency Industry, World Resources Institute.

de Ameida, A. T., F. J. T. E. Ferreira, et al. (2008). EUP Lot 11 Motors Preparatory Study, ISR - University of Coimbra.

de la Rue du Can, S. (2009). Residential and Transport Energy Use in India: Past Trends and Future Outlook, LBNL.

Delio, E. A., S. Lall, et al. (2009). Powering Up: The Investment Potential of Energy Service Companies in India, World Resources Institute.

EC (2007). ENER Lot 6 Stand by and Off Mode Losses Final Preparatory Study.

G. Pandian (2008). Progress of Standards and Labeling Programme in India. International Conference on Standby Power. New Delhi.

Garg, S. (2009). Standards and Labeling Program for Motors in India. MEPSA, Sydney, Australia.

International Energy Agency (2006). World Energy Outlook 2006, OECD.

Kumar, S., R. Kapoor, et al. (2010). Developing an Energy Conservation Building Code Implementation Strategy in India. ACEEE Summer Study.

Letschert, V. and M. McNeil (2007). Coping with Residential Electricity Demand in India's

Future - How Much Can Efficiency Achieve? ECEEE Summer Study. Côte d'Azur, France.

Mathur, A. (2011). National Mission on Enhanced Energy Efficiency. Energy Synergy Dialogue. World Energy Council.

McNeil, M. A., N. Bojda, et al. (2011). Business Case for Energy Efficiency in Support of Climate Change Mitigation, Economic and Societal Benefits in the United States, LBNL.

McNeil, M. A. and M. Iyer (2009). Progress towards Managing Residential Electricity Demand: Impacts of Standards and Labeling for Refrigerators and Air Conditioners in India. 5th International Conference on Energy Efficiency in Domestic Appliances and Lighting. Berlin.

McNeil, M. A., M. Iyer, et al. (2008). "Potential Benefits from Improved Energy Efficiency of Key Electrical Products: The Case of India." Energy Policy 36(9): 3467-3476. McNeil, M. A., V. E. Letschert, et al. (2008).

Global Potential of Energy Efficiency Standards and Labeling Programs, LBNL/CLASP.
Phadke, A., J. Sathaye, et al. (2005). Economic benefits of reducing Maharashtra's electricity shortage though end-use efficiency improvement, LBNL.
Prayas Energy Group (2010). "Energy Saving Potential In Indian Households From Improved Appliance Efficiency."
Price, L., M. D. Levine, et al. (2011). "Assessment of China's Energy-Saving and Emission-Reduction - Accomplishments and Opportunities During the 11th Five Year Plan." Energy Policy 39(4): 2165-2178.
Price, L., S. de la Rue du Can, et al. (2006). Sectoral Trends in Global Energy Use and Greenhouse Gas Emissions, Lawrence Berkeley Lab.
Sathaye, J., S. de la Rue du Can, et al. (2010). Strageties for Low Carbon Growth In India: Industry and Non-Residential Sectors, LBNL.
Singh, A. (2006). "Power sector reform in India: current issues and prospects." Energy Policy 34(16): 2480-2490.
Singh, D., G. Sant, et al. (2007). Review of Nashik Pilot CFL Program of Maharashtra State Electricity Distribution Co Ltd. Pune.
Tathagat, T. and M. Anand (2011). Impact Assesment for Refrigerator and Air-conditioner Voluntary Labeling Program in India, Collaborative Labeling and Appliance Standards Program.
The Energy Research Institute (TERI) (2008). TEDDY: TERI's Energy Data Directory and Yearbook 2007. New Delhi, TERI.
Weisbrod, G., M. Tribble, et al. (1998). Learning from India: Defining Profitable DSM and ESCo Programs for a Utility.
Zhou, N. (2010). "Overview of Current Energy Efficiency Policies in China." Energy Policy 38(11): 6439-6452.

Appendix – Assumptions and Sources for Appliances Cost-Benefit Calculation

Equipment	Lifetime	Discount Rate	Baseline UEC	Target UEC	Baseline Price	Target Price
Electric Equipment	years	years	kWh		$	$
Incandescent Lamps (Residential)	Letschert 2007	Assumption - Res. 10%	60 W IL 4hr/day	15 W CFL 4hr/day	Chinese Data	

Appendix (Continued)

Equipment	Lifetime	Discount Rate	Baseline UEC	Target UEC	Baseline Price	Target Price
Electric Equipment	years	years	kWh		$	$
Incandescent Lamps (Commercial)	Letschert 2007	Assumption - Com. 10%	60 W IL 8hr/day	15 W CFL 8hr/day		
Fluorescent Lamps (Residential)	Assumption	Assumption - Res. 10%	F40T12/ES Mag. 4 hr/day	Hi-perf T8 Elec. 4 hr/day		
Fluorescent Lamps (Commercial)	Assumption	Assumption - Com. 10%	F40T12/ES Mag. 8 hr/day	Hi-perf T8 Elec. 8 hr/day	Sathaye (2010)	
Refrigerators (Direct Cool)	McNeil 2008 (1)	Assumption - Res. 10%	4 Star	5 Star		
Refrigerators (Frost Free)	McNeil 2008 (1)	Assumption - Res. 10%	3 Star	5 Star	CLASP - EDS Impacts	
Air Conditioners (Residential)	Letschert 2007	Assumption - Res. 10%	2 Star Window, 3 Star Split	5 Star	CLASP - EDS Impacts	
Standby Power	Ecodesign Prep Study	Assumption - Res. 10%	Ecodesign Prep Study	Ecodesign Prep Study	Assumption	
Commercial Cooling	Sathaye (2010)	Assumption - Com. 10%	Data from WB Study	Eff. Imp from WB Study	Sathaye (2010)	
Motors						
10 HP (90%)	Revised downward from 15 in McNeil 2008 (1)	Assumption - Ind. 10%	Ecodesign - EPACT (EFF2)	Ecodesign - NEMA Premium (EFF1)	Garcia 2007 - U.S. 2003 prices for rough equivalent to EFF2 and EFF1	
50 HP (9%)						
100 HP (.9%)						
Distribution Transformers						
25 kVA (6%)	McNeil 2008 (1)	Assumption - Ind. 10%	McNeil 2008 (1) Star 1	McNeil 2008 (1) Star 5	McNeil 2008 (1) Star 1 McNeil	2008 (1) Star
60 kVA (25%)						
100 kVA (44%)						
160 kVA (3%)						
200 kVA (14%)						

[1] McNeil, M. A., M. Iyer, S. Meyers, V. E. Letschert and J. E. McMahon (2008). "Potential benefits from improved energy efficiency of key electrical products: The case of India." Energy Policy 36(9): 3467-3476

[2] McNeil, M.A., V.E. Letschert and S.A.de la Rue du Can (2008). Global Potential of Energy Efficiency Standards and Labeling Programs. LBNL 760E

[3] Letschert, Virginie E. and M.A.McNeil. 2007. "Coping with Residential Electricity Demand in India's Future – How Much Can Efficiency Achieve?" LBNL-63199, July 2007.

[4] Garcia (2007) Garcia, A. G. P., A. S. Szklo, R. Schaeffer and M. A. McNeil (2007). "Energy-Efficiency Standards for Electric Motors in Brazilian Industry". Energy Policy 35(6): 3424-3439. LBNL-62395

[5] Sathaye, J., S. de la Rue du Can, et al. (2010). Strategies for Low Carbon Growth In India: Industry and Non-Residential Sectors, LBNL -4557E

[6] McNeil, M. A. and M. Iyer (2007). Techno-Economic Analysis of Indian Draft Standard Levels for Room Air Conditioners.

[7] Tathagat, T. and M. Anand (2011). Impact Assessment for Refrigerator and Air-conditioner Voluntary Labeling Program in India, Collaborative Labeling and Appliance Standards Program.

End Notes

[1] India Ministry of New and Renewable Energy (India MNRE), "Union Cabinet approves National Mission for Enhanced Energy Efficiency Mission " Press Release, 24-June, 2010 20:40 IST, http://pib.nic.in/newsite/erelease.aspx?relid=62791

[2] India Ministry of Power (India MoP), 2010. "National Mission on Enhanced Energy Efficiency", http://www.india.gov.in/allimpfrms/alldocs/15659.pdf

[3] CLASP "Summary of S&L Information for India", http://www.clasponline.org/clasp.online.worldwide.php?countryinfo=93

[4] Bureau of Energy Efficiency (BEE), date not specified. "BACHAT LAMP YOJANA CDM based CFL scheme", Ministry of Power, Govt. of India. http://emt-india.com/BEE-BLY/BhachatLampYojna.pdf

[5] For simplicity the high efficiency scenario assumes 100% of the market will reach the target level in 2015, a structure that closely resembles minimum efficiency performance standards. In the later years of the forecast, the scenario is not highly sensitive to the details of the market transformation.

[6] BUENAS is described completely in McNeil, M.A., V.E. Letschert and S.A.De la Rue du Can (2008). *Global Potential of Energy Efficiency Standards and Labeling Programs*. LBNL 760E.

[7] More information about the LEAP platform may be found at http://www.energycommunity.org

[8] This depends somewhat on the lifetime of the product. For refrigerators we may assume a 15 - year lifetime, but some refrigerators may last 20 years, so the turnover of the stock may not be complete by 2030.

[9] Other metrics such as life cycle cost and payback period establish cost effectiveness, but are not easily compared across disparate technologies and end uses.

[10] This target is likely to be conservative, especially since the current labeling program should be ratcheted to this level in the next few years.

[11] Star rating market shares are Star 1 – 165, Star 2 – 39%, Star 3 – 31%, Star 4 – 3% and Star 5 – 11% according to *BEE-Verified Savings Report for 2009-2010* available at www.beeindia.in

[12] Motor market shares assumed to be IE1 – 67%, IE2 – 6% and the remaining motors below standards (Source: ICPCI).

[13] http://220.156.189.26:8080/beeLabel/Schedules/Schedule4-DistributionTransformer.pdf

[14] Source: ICPCI

[15] Available at http://indiabudget.nic.in/es2009-10/chapt2010/tab133.pdf

[16] India Census, available at http://www.thehindu.com/multimedia Census 2011517160a.pdf

[17] Industrial motors are the exception – we use U.S. prices for these because recent Indian prices were available for this important end use. We consider the use of U.S. prices to be conservative.

[18] It is possible to model, for example, the reduction of sales or fuel switching resulting from price increases associated with efficiency regulations. This effect is not captured in BUENAS.

[19] The exception is the phase-out of incandescent lamps, which begins in 2012 in the *Business Case*.

Index

A

access, 92, 116, 159
accounting, 22
acid, 44, 115, 157
ACs, 87
administrative support, 86
age, 18
agencies, 10
alternative energy, 47, 117, 159
assessment, 13, 19, 21, 91, 138
awareness, 9

B

banks, 123
barriers, 47, 73, 75
base, viii, ix, 2, 12, 15, 16, 17, 19, 40, 70, 74, 75, 78, 89, 92, 94, 96, 112, 126, 136, 139, 140, 141, 142, 143, 144, 152, 155
base year, 16, 19, 94, 96, 140, 142, 152
BEA, 22, 71
BEE, x, 126, 132, 133, 134, 135, 144, 145, 147, 163, 164
Beijing, 13, 91, 104, 117, 118, 138, 160
benefits, viii, 2, 4, 8, 11, 21, 32, 40, 41, 42, 44, 45, 79, 80, 82, 89, 97, 108, 112, 113, 115, 116, 127, 128, 130, 131, 136, 155, 156, 157, 158, 161, 162
bias, 47
blogs, 118
boilers, 27, 29, 33, 34, 35, 37, 70, 93, 100, 105
bottom-up, 15, 92, 139
building code, 132
businesses, vii, 1, 44

C

Cabinet, 23, 133, 163
campaigns, 9
capacity building, 133
carbon, viii, ix, 2, 4, 8, 9, 12, 17, 38, 42, 44, 47, 78, 79, 84, 87, 89, 94, 110, 115, 117, 126, 127, 131, 136, 141, 152, 156, 157, 159
carbon dioxide, 4, 8, 9, 17, 44, 79, 84, 87, 94, 115, 127, 131, 141, 157
CEC, 11, 71, 74
CEE, 11, 71
Census, 164
certificate, 132
China, 1, iii, v, vii, ix, 13, 74, 77, 78, 82, 83, 84, 85, 86, 87, 88, 89, 91, 92, 93, 96, 97, 98, 99, 100, 101, 102, 103, 104, 105, 111, 116, 117, 118, 119, 120, 122, 138, 160, 161
Chinese government, 79, 84, 87, 100, 104, 117
cities, viii, 78, 85

classes, 22, 23, 24, 25, 27, 29, 31, 35, 101
cleaning, 27
climate, 132, 134
CO2, 3, 5, 6, 38, 42, 45, 46, 79, 81, 82, 88, 110, 111, 112, 115, 116, 119, 127, 128, 129, 152, 153, 154, 157, 158
coal, 2, 3, 44, 45, 100, 115, 157
Code of Federal Regulations, 71
collaboration, 86, 87, 92, 133
color, 134
commercial, viii, ix, x, 7, 8, 9, 10, 16, 18, 19, 25, 28, 29, 31, 34, 35, 36, 40, 42, 45, 74, 78, 81, 85, 93, 96, 98, 99, 100, 103, 104, 105, 106, 110, 114, 116, 126, 134, 136, 142, 143, 144, 145, 146, 148, 158
commercial lighting, ix, 8, 36, 42, 78, 81, 85, 110, 114
community, 47, 117, 159
computer, 87
concreteness, 13, 90, 137
conditioning, x, 19, 25, 28, 81, 96, 111, 126, 129, 142, 145, 146, 154
configuration, 22, 98
Congress, 6, 10, 39
conservation, 42, 115, 132, 156
construction, ix, 22, 25, 31, 32, 35, 40, 78, 85, 88, 105, 108, 112, 134, 149, 150, 155
consumer goods, 44
consumers, viii, ix, 2, 3, 8, 9, 11, 12, 19, 21, 31, 32, 34, 40, 44, 46, 78, 79, 82, 87, 90, 103, 105, 106, 108, 112, 116, 126, 127, 130, 136, 137, 145, 148, 149, 155, 156, 158, 159
consumption, viii, ix, x, 2, 3, 9, 16, 19, 21, 24, 26, 27, 28, 46, 78, 84, 94, 97, 98, 99, 100, 106, 126, 133, 134, 135, 140, 145
contamination, 44, 115, 157
controversial, 24
cooking, 27
cooling, 9, 25, 27, 28, 34, 36, 98, 99, 105, 106, 130, 134, 145, 154
cooperation, 88
cost effectiveness, 74, 122, 147, 163
cost saving, 93
cost-benefit analysis, 12, 20, 21, 37, 89, 90, 109, 136, 137, 152
covering, 10, 35, 84, 89, 97
customers, 88, 104, 136

D

data collection, 97
database, 15, 92, 97, 139, 143
decision makers, 47, 117, 159
Department of Energy, 10, 14, 19, 20, 48, 71, 74, 91, 118, 138
developing countries, ix, 78, 85
diffusion, 16, 94, 140
distribution, x, 7, 18, 19, 29, 35, 37, 40, 81, 83, 95, 101, 109, 111, 114, 126, 130, 134, 136, 140, 141, 142, 143, 146, 147, 152, 154
district heating, 98, 105
domestic demand, 87
draft, 65, 66, 133
DSM, 11, 87, 88, 118, 133, 136, 161
dynamism, 26, 93, 140

E

economic savings potential, viii, ix, 2, 12, 78, 89, 126, 137
economic value, vii, viii, 2
efficiency level, 8, 12, 16, 20, 21, 31, 37, 41, 47, 82, 90, 97, 98, 105, 109, 114, 130, 137, 143, 146, 149, 152, 156
electricity, vii, viii, ix, x, 2, 3, 10, 11, 14, 21, 24, 25, 26, 27, 28, 29, 31, 37, 38, 44, 45, 46, 81, 84, 87, 88, 89, 91, 92, 101, 103, 109, 110, 111, 116, 126, 134, 135, 136, 138, 148, 150, 151, 152, 157, 161
electrodes, 28
electromagnetic, 114
emission, 81, 110, 129, 135, 154
employment, 96
energy conservation, 22, 101
energy efficiency, vii, viii, x, 2, 9, 10, 11, 13, 14, 15, 19, 21, 32, 40, 83, 85, 86, 87,

88, 89, 90, 91, 92, 101, 108, 112, 126, 131, 132, 133, 134, 135, 136, 138, 139, 145, 155, 162
Energy Independence and Security Act, 6, 10, 16, 24, 73
energy performance, vii, viii, ix, 1, 8, 9, 47, 77, 84, 85, 125, 131, 134
Energy Policy and Conservation Act, 9
energy prices, 5, 12, 31, 32, 81, 89, 90, 129, 136, 137
enforcement, 24, 85, 97
engineering, 21
Environmental Protection Agency, 10, 71, 118
environments, 28
EPA, 10, 44, 71, 115, 118, 157
EU, 145, 146
European Union, 145
evolution, 37, 109, 152
expenditures, 44

F

Federal Register, 71
financial, x, 8, 9, 10, 17, 32, 40, 41, 42, 44, 45, 82, 87, 89, 94, 108, 112, 113, 115, 116, 117, 126, 130, 141, 155, 156, 157, 158, 159
financial incentives, 9, 89
first generation, 9
fish, 44, 115, 157
flue gas, 25
force, 21
forecasting, 36
formula, 19, 96, 142
France, 160
frost, 134, 144, 149
funding, 88
funds, 133

G

GDP, 22, 79, 84, 86
GDP deflator, 22
Germany, 119
government budget, 88
governments, 87, 89
greenhouse, viii, ix, 4, 9, 10, 44, 45, 78, 79, 80, 85, 115, 116, 127, 128, 157, 158
greenhouse gas, viii, ix, 4, 9, 10, 44, 45, 78, 79, 80, 85, 115, 116, 127, 128, 157, 158
greenhouse gas emissions, viii, 4, 9, 10, 44, 45, 78, 79, 80, 85, 115, 116, 127, 128, 157, 158
grids, 4, 80, 128
Gross Domestic Product, 71
growth, viii, x, 9, 13, 16, 18, 44, 78, 85, 91, 92, 94, 95, 96, 117, 126, 130, 138, 140, 141, 152, 154, 159
growth rate, 152
Guangzhou, 87, 104
guidelines, 20

H

health, 12, 44, 89, 115, 136, 157
heat pumps, 27, 28, 81, 93, 98, 105, 111, 146
heat transfer, 25, 27
history, viii, 2, 9, 10, 11, 26, 74
homes, 27, 93, 99, 100, 130, 154
hotels, x, 126
household income, 36, 85, 92
household sector, x, 126
housing, 35
HVAC, ix, 7, 24, 40, 74, 78, 85

I

ID, 28
identification, 15, 92, 139
IEA, 26, 36, 72, 86, 100, 119, 145, 152
Impact Assessment, 163
improvements, 2, 8, 9, 11, 13, 14, 15, 21, 25, 26, 28, 31, 32, 33, 34, 41, 44, 46, 47, 74, 82, 83, 88, 90, 91, 92, 93, 108, 114, 117, 122, 130, 131, 138, 139, 140, 156, 159

income, viii, 78, 92
increased access, x, 126
Independence, 16, 39
India, 1, iii, v, vii, x, 14, 91, 118, 125, 126, 130, 131, 132, 133, 134, 135, 136, 138, 140, 143, 144, 145, 147, 148, 150, 152, 154, 158, 159, 160, 161, 162, 163, 164
Indian households, ix, 126, 158
Indian market, x, 126, 150, 159
induction, 86, 93, 100, 134
industries, 89, 133
industry, viii, 13, 78, 79, 84, 85, 91, 101, 118, 138
inefficiency, 29
inflation, 22, 104
infrastructure, 86
insulation, 6, 24, 25, 39
interest rates, 103, 148
International Energy Agency, 2, 46, 72, 119, 152, 160
investment(s), viii, ix, 2, 3, 4, 5, 8, 11, 12, 19, 20, 22, 40, 41, 45, 46, 47, 78, 79, 80, 82, 83, 85, 89, 90, 96, 97, 112, 113, 114, 116, 117, 126, 127, 129, 130, 132, 135, 136, 137, 142, 143, 155, 156, 158, 159
investors, 132
issues, 132, 161

J

Japan, 74
job creation, 44, 45

K

Kyoto Protocol, 135

L

labeling, 10, 12, 84, 86, 89, 101, 132, 133, 134, 137, 144, 145, 146, 163
lakes, 44, 115, 157
lead, 15, 92, 139
leadership, 87
leakage, 100, 147
learning, 12, 90, 137
LED, 33
legislation, 10, 11
life cycle, 74, 122, 163
lifetime, 18, 20, 37, 74, 95, 96, 97, 99, 100, 101, 106, 109, 122, 142, 143, 145, 146, 147, 149, 152, 163
light, 16, 28, 36, 87
local government, ix, 78, 85, 88
LPG, 134

M

magnet, 100
magnitude, 35, 44, 115, 157
majority, x, 20, 22, 126
management, 11, 12, 86, 87, 88, 90, 135, 137
manufacturing, 21
market penetration, 89
market share, 22, 25, 36, 82, 98, 111, 144, 164
mass, 33, 74, 75
materials, 13, 46, 91, 117, 138, 159
MBP, 60
median, 11
mercury, 3, 39, 44, 46, 79, 110, 115, 116, 157
Mercury, 50, 123
metal salts, 28
methanol, 100
methodology, 13, 47, 90, 92, 117, 135, 137, 159
middle class, x, 126
migration, viii, 78, 85
models, 10, 15, 18, 36, 81, 92, 96, 97, 111, 134, 139, 142
modifications, 15, 93, 140
multimedia, 164

N

National Research Council, 14, 91, 138

natural gas, 2, 3, 9, 25, 27, 29, 31, 37, 38, 42, 45, 100, 103, 105, 109, 110, 148, 151, 152
New Zealand, 119
normal distribution, 18, 95, 142

O

OECD, 72, 119, 160
oil, 9, 38, 110, 152
operating costs, 12, 89, 136
operations, 13, 91, 138
opportunities, viii, 2, 5, 9, 16, 20, 80, 117, 128, 159
ownership, viii, x, 16, 18, 36, 78, 85, 92, 94, 126, 140
ownership rates, ix, 18, 36, 78, 85, 92

P

parallel, 10
payback period, 74, 122, 163
PCA, 9
permit, 7, 15, 40, 93, 140
plants, 44, 73, 75, 115, 157
platform, 74, 122, 163
policy, 6, 15, 16, 17, 18, 21, 24, 37, 39, 40, 47, 86, 92, 94, 95, 97, 109, 113, 117, 132, 139, 140, 141, 142, 152, 155, 159
policymakers, 1, 8, 77, 84, 125, 131
pollutants, 39, 110
population, 15, 92, 139
portfolio, 88
power plants, 11, 44, 115, 157
present value, 8, 41, 82, 114, 130, 156
principles, 133
private investment, 135
private sector, 47, 117, 135, 159
procurement, 133
production costs, 22
project, 15, 92, 93, 133, 135, 139
prototype, 47, 117, 159
public investment, 41, 113, 156
pumps, 28, 98

R

R&D investments, 47, 117, 159
radio, 85
reality, vii, 2
recovery, 20, 96, 143
reform(s), 86, 122, 135, 161
regulations, 20, 21, 75, 88, 89, 123, 145, 146, 164
regulatory framework, 100
repair, 22, 74
replacement rate, 18
requirements, 29, 34
Residential, 4, 14, 15, 27, 33, 34, 36, 48, 50, 51, 52, 53, 55, 56, 57, 67, 68, 71, 72, 73, 79, 80, 81, 83, 93, 100, 102, 104, 106, 108, 111, 114, 118, 119, 120, 121, 122, 127, 131, 139, 147, 149, 150, 160, 161, 162, 163
restructuring, 135
retail, x, 11, 14, 22, 24, 87, 89, 91, 97, 105, 106, 126, 136, 138
retirement, 18, 30, 37, 95, 103, 109, 142, 148, 152
revenue, 11, 44
risk, 44, 115, 157
root, 21

S

saturation, ix, 78, 85
scaling, 19, 96, 142, 152
scope, x, 4, 80, 93, 126, 128, 133
security, 28
sensitivity, 148
shape, 18, 22, 95, 142
shortage, 87, 135, 161
showing, 24
social development, 132
society, 11, 89, 136
sodium, 86
spending, 11, 13, 44, 91, 138
stakeholders, vii, viii, ix, 1, 8, 77, 84, 125, 131

state(s), 10, 11, 104, 134, 135, 136, 148
stock, x, 15, 16, 17, 18, 19, 36, 37, 40, 74, 92, 94, 95, 109, 113, 122, 126, 139, 140, 141, 142, 152, 155, 163
storage, 25, 134
stoves, 100, 134
structure, 17, 35, 37, 74, 94, 109, 122, 140, 151, 152, 163
Subsidies, 87
subsidy, 87
substitution, 106, 149
suppliers, 87, 135
supply curve, 42, 115, 156
survival, 18, 95, 141
swelling, x, 126

T

target, 3, 8, 12, 15, 16, 21, 25, 31, 32, 37, 41, 46, 74, 75, 82, 86, 89, 90, 92, 93, 105, 108, 109, 114, 122, 130, 137, 139, 144, 145, 146, 147, 149, 152, 156, 163
tariff, 23, 74, 103, 148
tax incentive, 87
taxes, viii, ix, 2, 12, 47, 78, 89, 117, 126, 136, 159
technical support, 86
technologies, ix, 4, 9, 12, 15, 19, 36, 45, 46, 47, 74, 78, 79, 80, 84, 85, 89, 90, 92, 96, 97, 114, 116, 122, 127, 128, 137, 139, 142, 143, 146, 150, 158, 159, 163
tertiary sector, 96
time frame, 4, 8, 79, 84, 127, 131
top-down, 15, 92, 139
total energy, 3, 41, 46, 113, 155
total utility, 11
tracks, 18, 37, 95, 109, 141, 152
trade, 47, 117, 159
transformation, vii, viii, ix, 1, 3, 8, 12, 74, 77, 79, 84, 89, 122, 127, 131, 137, 163
transmission, 37, 38, 109, 110, 152
transport, 14, 91, 138
transportation, 4, 80, 128
triggers, 35
turnover, ix, 15, 74, 78, 85, 92, 122, 139, 163

U

U.S. economy, 44
UNFCCC, 135
United, v, vii, viii, 1, 2, 3, 9, 14, 19, 24, 25, 27, 36, 45, 46, 91, 138, 143, 148, 160
United States, v, vii, viii, 1, 2, 3, 9, 14, 19, 24, 27, 36, 45, 46, 91, 138, 143, 148, 160
urban, viii, 78, 88, 92
urbanization, 92

V

vacuum, 6, 34, 39
vehicles, ix, 78, 85
ventilation, 94

W

waste, 25, 27
waste heat, 25, 27
water, 6, 8, 9, 16, 21, 24, 25, 27, 34, 39, 41, 73, 74, 75, 82, 86, 93, 99, 100, 111, 134, 136
water heater, 6, 8, 9, 16, 21, 25, 27, 34, 39, 41, 73, 74, 82, 86, 93, 99, 100, 111, 134
websites, 97
White Paper, 84, 118
wholesale, 87
World Bank, 131, 143
worldwide, 163

Y

yield, 8, 18, 40, 82, 94, 112, 130, 141, 155, 156